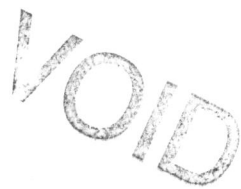

Library of
Davidson College

ELEMENTARY RADIATION PHYSICS

"How much physics will a doctor need?" Mach's answer was, "A doctor always needs exactly as much physics as he knows. Knowledge one does not possess one cannot use." Otto Blüh, "Ernst Mach as Teacher and Thinker," *Physics Today*, June 1967.

ELEMENTARY RADIATION PHYSICS

G. S. Hurst

Department of Physics and Astronomy
University of Kentucky
Lexington, Kentucky

J. E. Turner

Health Physics Division
Oak Ridge National Laboratory
Oak Ridge, Tennessee

JOHN WILEY & SONS, INC.
New York · London · Sydney · Toronto

539.72
H966e

72-1405

Copyright © 1970 by John Wiley & Sons, Inc.

All rights reserved. No part of this book may be reproduced by any means, nor transmitted. nor translated into a machine language without the written permission of the publisher.

10 9 8 7 6 5 4 3 2 1

Library of Congress Catalogue Card Number: 70-94921
SBN 471 42472 2
Printed in the United States of America

DR. KARL Z. MORGAN

in appreciation of his personal inspiration to us

Foreword

Everyone is continuously exposed to a certain amount of radiation—called background radiation or simply background—occurring in his environment. This background comes from sources such as (1) the small amounts of radioactive isotopes in rocks and in stone building materials, (2) radium in drinking water, (3) the natural potassium radioisotope $^{40}_{19}K$, (and others) in the body, (4) fallout from nuclear weapons tests, and (5) cosmic radiation. Cosmic rays, which were discovered in 1914, consist mainly of high-energy protons and other nuclei incident on the earth from outer space. Some of these primary rays penetrate to the earth's surface; some are absorbed in the atmosphere; and others interact with nuclei in the atmosphere to produce showers of secondary cosmic rays that reach the earth's surface.

The table below gives an analysis of the average distribution of radiation dose, based on its source, to a person living today in the United States. Approximately one-half of the exposure of the population results from man-made sources, and one-half results from naturally occurring sources. The largest man-made source is medical and dental X-rays. Radioactive fallout contributes about 5% of the total dose, and cosmic rays contribute about 10%. Local conditions vary from place to place, and the numbers given in the table are estimates of the average. In rocky areas, for example, background is high because of the presence of radioisotopes in rocks and radon gas in the air. Cosmic-ray intensity doubles in going from sea level to elevations of several thousand feet.

The development of modern medical practice and the arrival of the atomic age have made ionizing radiation a major factor in the environment of today's civilization. It is natural that a gap exists between the knowledge, held by relatively few specialists, and its dissemination as a part of the liberal arts or the more specialized premedical curriculum. This book is an attempt to fill the gap. In such a new, active, and exciting field, one is impressed both by what has been discovered and what yet remains unknown. The benefits gained thus far charge us with the responsibility of using wisely the knowledge we have discovered.

Sources of Radiation Dose to Persons in the United States[a]

Source	Approximate Percent Dose
Man-made	
Diagnostic X-ray	40
Theraputic X-ray	5
Radioactive fallout	5
Others	1
Subtotal	51
Natural	
Terrestial radiation, External to body	27
Cosmic rays	12
Radioisotopes, internal to body	10
Subtotal	49
Total	100

[a]Based on *Radiation Protection Criteria and Standards: Their Basis and Use*, Summary-Analysis of Hearings before The Special Subcommittee on Radiation of the Joint Committee on Atomic Energy, Congress of the United States. U.S. Government Printing Office, Washington, D. C., October 1960.

Preface

Radiation physics is much more than a subject of interest to a specialized group of scientists. Every living thing is exposed to a background of ionizing radiation originating from naturally occurring radioactive elements, cosmic rays, and fallout from nuclear explosions. A large fraction of the world's population is also exposed to X-rays from medical and dental sources. An unfortunate few are involved in serious radiation accidents. Because radiation affects all of us, we should have some knowledge of it—just as we should become acquainted with all other major elements of our environment. This book is not written to alarm or to pacify, but simply to inform.

The material is presented at a level that presupposes little knowledge of physics: a course in general physics at the high school or college level is an adequate prerequisite. This book explains in elementary terms what radiation is, how it interacts with matter, and how it is measured. It describes some of the harmful effects and beneficial uses of radiation. We have stressed the basic principles of radiation physics instead of assembling a collection of the more publicized radiation phenomena. A limited knowledge of mathematics is needed to follow the exposition of the principles and to work the problems illustrating them.

The material is designed for use in college-level physics courses and as short courses in radiation physics for students of engineering, premedicine, predentistry, and biology. We believe that it especially fills a need for teaching material for beginning courses in health physics, radiological health specialties, and other professional training programs.

We acknowledge with appreciation the encouragement given us by students and faculty members. Suggestions from M. T. Mc Ellistrem, W. C. Demarcus, J. A. Sayeg and R. E. Knight and the assistance of M. Fast in preparing lecture demonstrations and in making available the interesting historical material on X-rays merit special mention. We thank

Kathryn E. Lockridge and James D. Cape, Headquarters, U.S. Atomic Energy Commission, for furnishing much of the background material and photographs used in the preparation of several chapters. We thank R. B. Vora and Don Hurst for their help in reading proof.

We are grateful to Betty Hurst and Renate Turner for encouragement and, especially, to Renate Turner for typing the manuscript and correspondence.

Lexington, Kentucky, 1969

G. S. Hurst

J. E. Turner

Contents

CHAPTER

1 Electromagnetic Radiation and Atomic Structure	1
2 Radiation and the Nucleus	20
3 Interactions of Charged Particles with Matter	37
4 Interactions of X-rays and Gamma Rays with Matter	47
5 Interaction of Neutrons with Matter	61
6 Absorbed Energy and Its Measurement	69
7 Radiation Dosimetry	81
8 Biological effects and the Control of Radiation Exposure	95
9 X-Ray Technology	118
10 Applications of Radiation and Radioisotopes	132

APPENDIX

A Historical Outline	153
B Symbols and Abbreviations	155
C Physical Constants	157
D Conversion factors	158
E Values of the Exponential Function	159
Index	161

ELEMENTARY RADIATION PHYSICS

CHAPTER ONE

Electromagnetic Radiation and Atomic Structure

> The discovery of these beautiful and simple laws concerning the line spectra of the elements has naturally resulted in many attempts at a theoretical explanation. Such attempts are very alluring because the simplicity of the spectral laws and the exceptional accuracy with which they apply appear to promise that the correct explanation will be very simple and will give valuable information about the properties of matter.
>
> *Niels Bohr*

1-1 INTRODUCTION

The accidental discovery of X-rays in 1895 by Roentgen marked the beginning of man's knowledge of a phenomenon called ionizing radiation. Ionizing radiation differs from other kinds of radiation, such as radio waves and visible light, in having the ability to knock electrons out of matter, leaving charged atoms or ions. Although its presence is not detected directly by human senses, it causes photographic film to darken, gases to conduct electricity, and certain materials to scintillate or fluoresce; and it produces other effects that we shall describe later.

Additional kinds of ionizing radiation were subsequently found, including those from radioactive materials and cosmic rays. The emission of radiation from matter and the interaction of radiation with matter are the central topics of radiation physics. To discuss either process requires an understanding of the structure of matter. Therefore, in these first two chapters we present a review of atomic and nuclear structure and its connection with the emission of light, X-rays, gamma rays, and particles from radioactive materials. In subsequent chapters the interaction of radiation with matter, its detection, measurement, and some of its applications are described.

1-2 ATOMIC STRUCTURE OF MATTER

The smallest unit of a chemical element is the atom. In chemical reactions, atoms maintain their identity and join together in discrete numbers and in definite geometrical patterns to form molecules. The mixing of hydrochloric acid and sodium hydroxide, for example, produces salt and water. This reaction is represented by writing

$$HCl + NaOH \rightarrow NaCl + H_2O$$

in which the symbols stand for individual atoms. Modern atomic theory had its beginning in the early 19th century when Dalton studied quantitatively the combining of different elements and made the first measurements of atomic weight. The concept of the combining of individual atoms, having characteristic atomic weights, accounts for the observed laws of constant proportions and multiple proportions.

After the discovery of the electron at the end of the 19th century, Thomson proposed the "plum pudding" picture of the atom shown in Fig. 1-1. In the Thomson model electrons, which are negatively charged and have 1/1836 the mass of the lightest atom (hydrogen), are emersed in the atom's positive charge. This mixture of electrical charges of two signs was thought to be distributed throughout the atomic volume.

The inner structure of atoms was revealed, however, only after the discovery of ionizing rays, which were used as atomic probes. A high-speed ray sent into an atom is scattered and comes out in a manner that depends

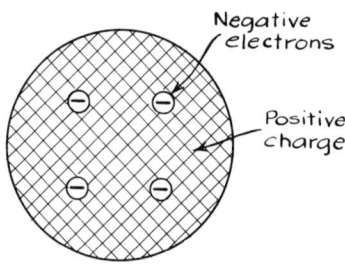

Fig. 1-1 Thomson "plum pudding" model of the atom (c. 1900).

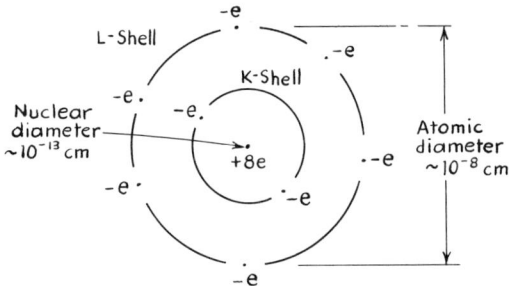

Fig. 1-2 Planetary model of the atom as proposed by Rutherford in 1911. The atom shown here, oxygen, has a total of 8 electrons, each of charge $-e = -1.602 \times 10^{-19}$ coulomb, in orbit about a nucleus of charge $+8e$ (nuclear charge = atomic number of element = 8 for oxygen). The nucleus of oxygen is about 30,000 times more massive than one electron, although the nuclear diameter is only about 10^{-5} that of the atom. The innermost shell, called the K-shell, of an element is occupied by at most two electrons. The next, L-shell, is occupied by at most 8 electrons. Elements of higher atomic number have filled K, L, M, etc. shells. The outermost shell of a neutral atom is filled only in the case of the noble, or inert, gases (atomic numbers = 2, 10, 18, 36, 54, 86). When two atoms are brought close together their outermost (valence) electrons interact most strongly. The number of electrons in the outermost shell thus determines the gross chemical properties of an element and its position in the periodic system.

on specific properties of the atom, such as the distributions of its charge and mass. By 1911 the alpha-particle scattering experiments of Rutherford and co-workers had shown that the positive atomic charge resides entirely in a tiny, massive nucleus at the center of the atom. The negative electrons orbit about the nucleus, much as the planets orbit about the sun in the solar system (Fig. 1-2). The number of positive electronic units of charge carried by the nucleus is the element's atomic number, which determines its position in the periodic system of the chemical elements.

This planetary picture of the atom was put on a quantitative basis in 1913 by Bohr's theory of the hydrogen atom and single-electron ions, which is discussed in Section 1-5. Although Bohr's theory later proved to be inadequate for treating more complicated systems, it provided an essential step toward the discovery of the highly successful quantum mechanics by Heisenberg and by Schroedinger in 1925–1926. In the modern view, electron clouds replace the orbiting electrons of the Rutherford-Bohr model (Section 1-7).

1-3 ATOMS, ISOTOPES, AND THE PERIODIC TABLE

The lightest atom found in nature—that of the element hydrogen—consists of a single electron in orbit about a nuclear particle (called a proton) of unit positive electronic charge. This atom is denoted by the symbol $_1^1H$, where the subscript gives the atomic number and the superscript the atomic mass number, which for this atom is unity. The second lightest atom is also hydrogen and has a weight of almost exactly two units, $_1^2H$. The nucleus in this case consists of the single proton bound closely to a neutral particle (called a neutron) of slightly greater mass than the proton. The neutron will be discussed in Chapter 2. Different varieties of atoms of a chemical element, having the same atomic number but different weights, are called isotopes of the element. The isotope $_1^2H$ is called heavy hydrogen, or deuterium. A third isotope of hydrogen with three units of weight, $_1^3H$, called tritium, is also found in relatively small amounts in nature. Two isotopes of the next element, helium, in the periodic system, $_2^3He$ and $_2^4He$, are found in nature.

Since different isotopes of an element have the same number of charges and orbital electron structure, they exhibit similar chemical properties. Isotope separation cannot be accomplished by chemical methods. Instead, methods that depend on physical differences between the atoms of different masses are used. Practical methods of separation include gaseous diffusion, centrifugation, and bending of particle paths in magnetic fields.

All isotopes have nearly integral values of weight. On the unified scale of atomic weights the isotope $_6^{12}C$ is given an atomic weight of exactly 12.

TABLE 1-1 Data for Some Elements

Element	Symbol	Atomic Number	Atomic Weight (Unified Scale)
Hydrogen	H	1	1.0080
Boron	B	5	10.811
Carbon	C	6	12.011
Nitrogen	N	7	14.007
Oxygen	O	8	15.999
Sodium	Na	11	22.990
Aluminum	Al	13	26.982
Chlorine	Cl	17	35.453
Potassium	K	19	39.102
Calcium	Ca	20	40.08
Iron	Fe	26	55.847
Silver	Ag	47	107.870
Gold	Au	79	196.967
Lead	Pb	82	207.19
Radon	Rn	86	222
Radium	Ra	88	226
Uranium	U	92	238.03

Over 300 isotopes occur among the 90 natural elements. About 20 elements have each a single natural isotope, their atomic weights being almost integral. Other elements occur as mixtures of isotopes and have nonintegral atomic weights. Chlorine, for example, which occurs with two isotopes, $^{35}_{17}\text{Cl}$ and $^{37}_{17}\text{Cl}$, has an atomic weight of 35.5. As found in nature, this element consists of approximately 75% $^{35}_{17}\text{Cl}$ and 25% $^{37}_{17}\text{Cl}$ (Problem 1). The atomic numbers and weights of some elements are given in Table 1-1.

The gram atomic weight of an element is defined as the amount that has a mass in grams equal to the element's atomic weight. We see from Table 1-1, for example, that the gram atomic weight of hydrogen is 1.0080 g and that of oxygen, 15.999 g. The same number of atoms are contained in a gram atomic weight of any element. This number, which is called Avogadro's number, is 6.023×10^{23}. Knowing Avogadro's number, we can calculate weights of single atoms from the atomic weights. When more than one isotope exists, this weight is the average of the various isotopes. The average weight of a single atom of oxygen found in nature, for example, is $15.999/6.023 \times 10^{23} = 2.66 \times 10^{-23}$ g.

1-4 QUANTUM THEORY OF ELECTROMAGNETIC RADIATION

According to Maxwell's classical theory of electromagnetism, energy is emitted in the form of electromagnetic waves when an electric charge is accelerated. A charge oscillating with simple harmonic motion, for example, emits waves having the frequency of the motion. The oscillatory motion of an electric current in an antenna is used to produce radio waves; the more abrupt changes in the distributions of charges in an atom give rise to higher frequency waves (e.g., visible and ultraviolet light). The range of frequencies and wavelengths in the spectrum of electromagnetic radiation is discussed in Section 1-10.

The wave nature of electromagnetic radiation accounts for its observed diffraction and interference. Its velocity $c = 3 \times 10^8$ m/sec (in vacuum), wavelength λ and frequency of vibration ν are related by the formula

$$\lambda \nu = c \tag{1-1}$$

Under some conditions, which we shall describe later (Section 4-3 on the photoelectric effect), electromagnetic radiation also exhibits the "nonwave" characteristics of discrete, massless packets of energy, called photons, or light quanta. The relationship between the photon energy E and the frequency ν of the radiation was discovered by Einstein:

$$E = h\nu \tag{1-2}$$

6 Electromagnetic Radiation and Atomic Structure

Here h, like c, is a fundamental constant of nature, called Planck's constant ($h = 6.625 \times 10^{-34}$ j-sec). Modern quantum mechanics accounts for the dual wave-particle character of electromagnetic radiation found experimentally. Quantum mechanics also describes the observed wave characteristics of the electron and other particles of matter (Section 1-6).

1-5 BOHR THEORY OF THE HYDROGEN ATOM

Since, according to the laws of classical mechanics, an accelerated charge emits electromagnetic radiation, the atom pictured in Fig. 1-2 would be unstable. The orbiting electrons, experiencing a centripetal acceleration, would continuously lose energy through radiation and thereby spiral into the nucleus.

Bohr introduced two nonclassical postulates in his theory of the hydrogen atom. He assumed, first, that the electron moves without radiating only in certain discrete orbits about the nucleus. Second, he assumed that the electron emits (or absorbs) a photon of light only when it simultaneously jumps from one of these orbits to another of lower (or higher) energy. In either case, the energy of the photon is equal to the difference in the electron's energies in the two orbits. The hydrogen atom is thus characterized by its ability to emit and absorb only certain wavelengths of electromagnetic radiation. These wavelengths are described as the hydrogen atom spectrum.

Bohr found that the particular electronic energy levels that lead to the observed hydrogen spectrum result when the angular momentum of the electron about the nucleus is required to be an integral multiple of $\hbar =$ Planck's constant divided by 2π:

$$\left.\begin{array}{r}\text{Angular momentum of}\\ \text{electron about nucleus}\end{array}\right\} = \frac{nh}{2\pi} = n\hbar \qquad (n = 1, 2, 3, \ldots) \qquad (1\text{-}3)$$

The resulting levels of energy E_n are found to be inversely proportional to the square of the quantum number n. To show this, we represent in Fig. 1-3 the electron of mass m, charge $-e$, and constant speed v in uniform circular motion at a distance r from a nucleus of charge Ze. The condition (1-3) then implies that

$$mvr = n\hbar \qquad (1\text{-}4)$$

Equating the centripetal force mv^2/r on the electron to the electrostatic, or Coulomb, force KZe^2/r^2 where $K = 9 \times 10^9$ new-m²/coul², we find that

$$r = \frac{KZe^2}{mv^2} \qquad (1\text{-}5)$$

Eliminating v between Eqs. 1-4 and 1-5 gives the allowed orbital radii

$$r = \frac{n^2\hbar^2}{KZe^2m} = 0.529 \times 10^{-10} \frac{n^2}{Z} \text{ m} \qquad (1\text{-}6)$$

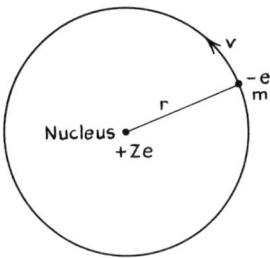

The radius of the lowest orbit ($n = 1$) in hydrogen ($Z = 1$), 0.529×10^{-10} m, is called the Bohr radius.

Eliminating r between Eqs. 1-4 and 1-5 gives the possible orbital speeds of the electron

Fig. 1-3 Bohr picture of single electron in orbit about nucleus (1913).

$$v = \frac{KZe^2}{n\hbar} = 2.19 \times 10^6 \frac{Z}{n} \text{ m/sec} \qquad (1\text{-}7)$$

With $Z = n = 1$, we find that the electron's speed in the state of lowest energy (ground state) of hydrogen is close to 1/137 that of light.

The allowed values of the electron's kinetic and potential energies are

$$KE = \tfrac{1}{2}mv^2 = \frac{K^2Z^2e^4m}{2n^2\hbar^2}$$

and

$$PE = -\frac{KZe^2}{r} = -\frac{K^2Z^2e^4m}{n^2\hbar^2}$$

The allowed total energies are given by

$$\begin{aligned} E_n = KE + PE &= -\frac{K^2Z^2e^4m}{2n^2\hbar^2} = -2.18 \times 10^{-18} \frac{Z^2}{n^2} \text{j} \\ &= -\frac{13.6Z^2}{n^2} \text{ eV}^1 \end{aligned} \qquad (1\text{-}8)$$

In the normal, or ground, state ($n = 1$) of hydrogen ($Z = 1$) the electron has its lowest energy, -13.6 eV. Energy (13.6 eV) is thus required to ionize the hydrogen atom. The electronic energy in the ground state of the single-electron helium ($Z = 2$) ion, He$^+$, is -54.4 eV, the negative of which is the energy required to obtain He^{++} from He$^+$.

The energy levels for atomic hydrogen given by Eq. 1-8 are shown in

---

[1]The unit of energy often used in atomic physics, the electron volt (eV), is defined as the change in the energy of an electron when it moves freely through a potential difference $V = 1$ volt $= 1$ joule/coulomb. Thus, in MKS units, $1 \text{ eV} = 1$ (joule/coulomb) $\times 1.602 \times 10^{-19}$ coulomb $= 1.602 \times 10^{-19}$ joule.

8 Electromagnetic Radiation and Atomic Structure

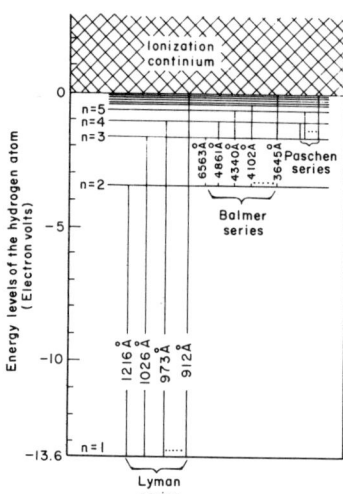

Fig. 1-4 Energy level diagram for atomic hydrogen.

Fig. 1-4. The internal changes in the hydrogen atom that lead to its observed emission spectrum are the possible electron jumps from one orbit to another of lower energy, nearer the nucleus. Passing an electric current through a tube containing hydrogen gas excites atoms from their ground state ($n = 1$) to states of higher quantum number ($n > 1$). The electron will be excited to one of the discrete energy states or, if it receives more than 13.6 eV of energy, the atom will become ionized. The potential energy of the free electron is then zero, and its kinetic energy can vary over a continuous range of positive values. The energy 13.6 eV is called the ionization potential of hydrogen.

In returning to the ground state, an electron may make a single jump toward the nucleus from a higher orbit or from the ionization continuum or it may cascade through a series of intermediate orbits. A photon of light is emitted in each jump, and the spectrum of light characteristic of hydrogen is given off by the gas atoms in the tube. The average time spent by an electron in an excited state is called the lifetime of the state. Typically, the lifetimes of excited atomic states are $\sim 10^{-8}$ sec.

The wavelengths of the emitted photons can be calculated from Bohr's theory. For example, the energy of the photon emitted in the lowest-energy transition in the Balmer series ($n = 3 \rightarrow n = 2$) is, according to Bohr's second postulate, equal to the energy difference between these two levels. We have, from Eq. 1-8, $E_2 = -13.6/(2)^2 = -3.40$ eV and $E_3 = -13.6/(3)^2 = -1.51$ eV. The energy of the photon emitted in this transition is thus

$$E_3 - E_2 = -1.51 - (-3.40) = 1.89 \text{ eV}$$
$$= 1.89 \text{ eV} \times (1.60 \times 10^{-19} \text{ j/eV})$$
$$= 3.02 \times 10^{-19} \text{ j}$$

The frequency of the light can be found from the Einstein relation (1-2):

$$\nu = \frac{E}{h} = \frac{3.02 \times 10^{-19} \text{ j}}{6.63 \times 10^{-34} \text{ j-sec}} = 4.56 \times 10^{14}/\text{sec}$$

The wavelength is, by Eq. 1-1,

$$\lambda = \frac{c}{\nu} = \frac{3 \times 10^8 \text{ m/sec}}{4.56 \times 10^{14}/\text{sec}} = 6.58 \times 10^{-7} \text{ m} = 6580 \text{ Å},$$

where the wavelength unit, 1 angstrom = 1 Å = 10^{-10} m, has been introduced. (More accurate calculation gives λ = 6563 Å, as observed). This light appears red in color. Other photons in the Balmer series are in the region of visible light (4000–6500 Å). The Lyman series, of shorter wavelength, is ultraviolet and the other series, of longer wavelength, are infrared.

While Bohr's theory accounted for the structure and electromagnetic radiation emitted and absorbed by the hydrogen atom (and single-electron ions, He^+, Li^{++}, etc.), it was only after the discovery of quantum mechanics in 1925–1926 that a satisfactory theory of more complex atoms and molecules was developed. Nevertheless, the break with the traditional ideas of classical mechanics in Bohr's theory gives us physical insight into the modern view of atomic structure. Atoms of different chemical elements have different electron energies, and each element has its own characteristic emission and absorption spectrum for electromagnetic radiation. More complex structures, like molecules of chemical compounds, have, in addition to their spectra from electronic transitions, characteristic spectra associated with the vibrational and rotational motion of the atomic nuclei relative to one another. Differences in the energies of these molecular motions correspond to the region of infrared light. Transitions within the atomic nucleus itself cause the emission of photons of very short wavelengths, called gamma rays. Whereas transitions in electronic and molecular motions determine the spectra of elements and their chemical compounds, gamma-ray photons have wavelengths characteristic of the isotopes present in a sample. Gamma-ray spectroscopy is discussed in the next chapter.

1-6 DE BROGLIE WAVES

In 1924 de Broglie proposed that the dual wave-particle character of photons might also be exhibited by particles of matter. The momentum p of a photon is given by the ratio of its energy E and its speed c. By means of Eqs. 1-1 and 1-2, we may write

$$P = \frac{E}{c} = \frac{h\nu}{c} = \frac{h}{\lambda} \qquad (1\text{-}9)$$

De Broglie suggested that the momentum and wavelength of a particle are related in the same way as those of a photon: $\lambda = h/p$. Since the momentum of a particle is $p = mv$, where m and v are its mass and velocity, the de Broglie wavelength of the particle is

$$\lambda = \frac{h}{mv} \qquad (1\text{-}10)$$

Equations 1-9 and 1-10 are valid in relativistic, as well as nonrelativistic, mechanics. Nonrelativistically, since the kinetic energy of a particle is given by $E = \frac{1}{2}mv^2$, we may substitute $v = \sqrt{2E/m}$ into Eq. 1-10 and write

$$\lambda = \frac{h}{\sqrt{2Em}} \qquad (1\text{-}11)$$

For an electron of energy E, expressed in electron volts, we find that the wavelength in angstroms (1 Å = 10^{-10} m) is given by

$$\lambda = \frac{12.3}{\sqrt{E(\text{eV})}} \text{ Å} \qquad (1\text{-}12)$$

An electron accelerated from rest through a potential difference of 100 V, for example, has a wavelength of 1.23 Å.

The wave nature of the electron and other particles of matter and the validity of Eq. 1-10 have been observed in the laboratory: for example in the diffraction of electrons and neutrons by crystals, and in the electron microscope. The resolution of a microscope is limited, in principle, by diffraction and interference to distinguishing objects no closer together (approximately) than the wavelength of the agent used for observation. Compared to the shortest visible optical wavelengths (~ 4000 Å), electron wavelengths (e.g., 0.123 Å when the energy of the electron is 10^4 eV) are many orders of magnitude smaller. Electron microscopes thus have much

1-7 DE BROGLIE WAVES AND ELECTRON CLOUDS

The de Broglie wavelength of an electron over a wide energy range, as described by Eq. 1-12, is comparable to atomic dimensions, and the wave nature of this particle plays an essential role in atomic physics. For example, Bohr's angular momentum quantization rule can be visualized with the help of Eq. 1-10. From Eq. 1-4, we have ($\hbar = h/2\pi$)

$$2\pi r = n\frac{h}{mv}$$

Applying Eq. 1-10, we see that

$$2\pi r = n\lambda \tag{1-13}$$

It follows that the circumferences of the orbits in hydrogen are integral numbers of electronic wavelengths.

In place of the classical concept of an electron as a point charge, occupying a given position at a definite time in its orbit about the nucleus, quantum mechanics represents it as a standing wave of charge density in the region of the classical orbit. Standing waves occur in connection with other phenomena, a simple example being the vibrating string represented in Fig. 1-5. With fixed length L the string can perform standing wave vibrations with only certain discrete wavelengths. As seen from Fig. 1-5, the wavelength λ, which is the distance between every other node, or point of zero displacement, must be restricted to values so that $L = n\lambda/2$, where $n = 1, 2, 3, \ldots$. In an analogous way, Eq. 1-13 describes the discrete states in the hydrogen atom. With $n = 2$, for example, the standing-wave electron cloud surrounds an orbit having a

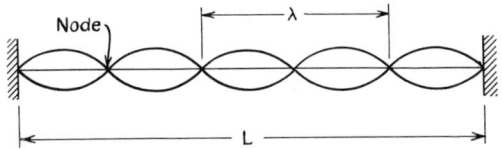

Fig. 1-5 Standing wave pattern with vibrating string.

12   Electromagnetic Radiation and Atomic Structure

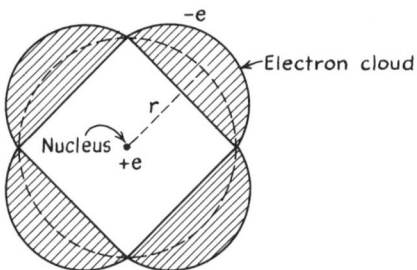

Fig. 1-6 Modern quantum mechanical picture of the hydrogen atom in $n = 2$ state.

circumference of two wavelengths (four nodes), as shown schematically in Fig. 1-6. Because of its wave nature, the position of the electron is unspecified within the dimensions of its wave packet.

1-8 X-RAYS

Roentgen, who in 1901 received the first Nobel prize in physics, discovered X-rays unexpectedly in experiments he was conducting in 1895. Studying the light emission produced when electricity is passed through a tube of gas at low pressure, he noticed that a paper screen coated with a fluorescent material glowed in the vicinity of the tube when it operated. The conditions of the discovery are shown schematically in Fig. 1-7. Under the influence of a high potential difference, which in modern equipment may be hundreds of thousands of volts or more, an

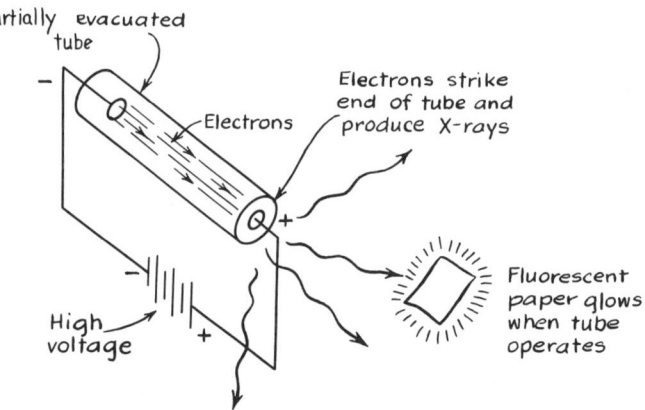

Fig. 1-7 Generation of X-rays in low-pressure gas tube.

electric current flows through the tube. This current consists of a stream of electrons that pass from the tube's negative terminal (cathode) to its positive terminal (anode). Electrons striking the anode are brought sharply to rest. Electromagnetic theory shows that the deceleration of an electric charge causes it to radiate energy, in this case in the form of electromagnetic radiation of very short wavelength, called X-rays.

Roentgen correctly attributed the fluorescence of the paper screen in Fig. 1-7 to the passage of a previously unknown agent, X-rays, from the tube to the screen. The rays were found to originate at the position where the electron beam struck the anode. The screen fluoresced even when located several meters from the tube, indicating that X-rays are able to pass through a considerable amount of air. Books and blocks of wood were found to be more transparent to the rays than were denser materials such as tin and lead.

Early observations showed that X-rays affect photographic film and ionize atoms. The latter property accounts for the fact that X-rays cause the discharge of an electrically charged object: nearby ions of air produced by X-rays are attracted to the object and neutralize its charge. These effects on matter—fluorescence, chemical change in photographic film, and ionization—are utilized in devices for detecting and measuring the properties of X-rays, as we shall discuss in later chapters.

X-rays are not deflected when they travel through electric and magnetic fields. Since the path of a charged particle is curved in these fields except under special conditions, this finding indicates that X-rays, like other kinds of electromagnetic radiation, are electrically neutral. Their lack of charge accounts in part for their ability to penetrate matter to great depths (Chapter 4).

The use of X-rays for diagnosis in medical science was recognized at once after the initial discovery. They are absorbed in different degrees by the light elements in soft tissue and the heavier elements in denser bone. Roentgen succeeded in photographing bones in the human body with X-rays. (See Fig. 9-1).

Early workers with X-rays discovered that exposure of the body to the radiation can produce burns, ranging in severity from mild reddening of the skin (erythema) to severe damage. Later it was found that X-rays also produce genetic changes in living systems. At low levels of exposure no observable damage may be produced, although the rays interact with atoms in the cellular environment. The administration of X-rays to a patient, therefore, implies an affirmative judgment—or, at least, a tacit assumption—that the expected benefit to be gained from their use outweighs the damage or risk to an exposed patient and to attending personnel. The levels of allowed exposure of personnel to radiation of all kinds in

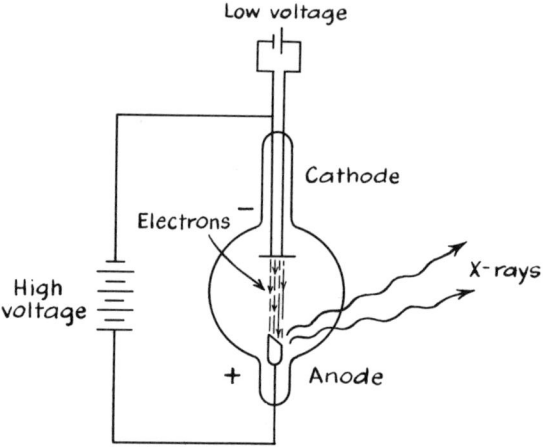

Fig. 1-8 Modern X-ray tube.

government operated atomic energy programs is based on a recognition of the need to balance risks and benefits (Section 8-9).

Figure 1-8 shows the basic design of a modern X-ray generating tube, which operates under high vacuum. A copious supply of electrons is obtained by heating a metal cathode with a current from a low-voltage supply. The electrons "boil" out of the hot cathode, and are accelerated to high energies in an intense electric field due to a high-voltage source applied between the cathode and anode. High accelerating voltages give "hard" X-rays, that is, those of highest energy and greatest penetrating ability in matter. X-ray tube anodes are often made of tungsten, which has a high melting point, and may be water-cooled by auxillary equipment. The spectrum (quality) of the X-rays is controlled by varying the high voltage; the intensity is independently controlled by means of the low voltage. More information on the development of X-ray tubes is given in Section 9-1.

The relative intensities of X-rays at various wavelengths – the X-ray spectrum – from a typical tube is shown in Fig. 1-9. There are two important features of the spectrum. First, a number of peaks occur at certain wavelengths λ_α, λ_β, ..., depending on the element of which the anode is made. This part of the radiation is called the characteristic spectrum. These photons are emitted as the result of electronic transitions within the anode atoms, like those illustrated for hydrogen in Fig. 1-4. Whereas the photons from hydrogen are in the visible or near-visible region, those from the inner shells of the high atomic number X-ray tube anode are more energetic and hence have shorter wavelengths. Second, there is a

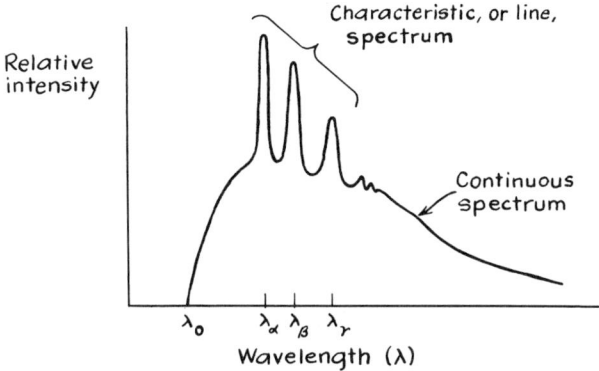

Fig. 1-9 Spectrum of X-rays from a tube.

continuous spectrum over a range of wavelengths above a minimum value λ_0 determined by the value of the accelerating voltage. These photons are emitted as the result of the decelerations of electrons striking the target near its nucleus. The maximum photon energy in the continuous spectrum is equal to the energy of the electrons accelerated by the high voltage (cf. Problem 33).

The shorter-wavelength (harder) rays are most useful in making medical X-ray pictures. Because X-rays damage tissue, it is important that the longer wavelength region of soft rays, which do not contribute as much to picture definition, be filtered out of the beam from an X-ray machine. Usually, a thin piece of aluminum placed in the path of the beam provides adequate filtration.

1-9 HEISENBERG UNCERTAINTY PRINCIPLE AND LIFETIME OF EXCITED STATES

As we mentioned at the end of Section 1-7, the exact position of an electron is not specified within the extent of its wave packet. Quantum mechanics shows that the probability of finding an electron in a small volume in space is proportional to the square of the amplitude of the wave packet in that volume.

The wave nature of the electron accounts for the properties of chemical bonds, and for many characteristics of matter not explained by the laws of classical physics. The physical significance of the wave properties of the electron and other particles is expressed by the uncertainty principle. Heisenberg showed that the product of the uncertainty, Δx, in our knowledge of the position of a particle in any direction and the uncertainty,

Δp_x, in our knowledge of its momentum in that direction must always be greater than, or — under optimum conditions — equal to, Planck's constant:

$$\Delta x \cdot \Delta p_x \geqq h \tag{1-14}$$

In contrast to classical physics, the position and momentum of a particle can never be known *simultaneously* with arbitrarily small precision. Accurate measurement of the position of a particle can be made only with corresponding uncertainty in the knowledge of its momentum, and vice versa.

In its general form, the Heisenberg uncertainty principle relates a number of pairs of observable quantities. Applied to the energy of atomic states, the uncertainty principle gives

$$\Delta E \cdot \Delta t \geqq h \tag{1-15}$$

where ΔE is the uncertainty in the knowledge of the energy of the state and Δt is its lifetime. The existence of this relation implies that every atomic energy level, such as those shown in Fig. 1-4, has a finite width, called the natural line breadth. (The characteristic photons emitted by an element are most often observed as a series of lines — called spectral lines — in a spectroscope). We can estimate the natural line breadth of an atomic state having a lifetime $\Delta t \sim 10^{-8}$ sec:

$$\Delta E \sim \frac{h}{\Delta t} \sim \frac{6.63 \times 10^{-34} \text{j-sec}}{10^{-8} \text{ sec}} \sim 6.63 \times 10^{-26} \text{j} \times \frac{1 \text{ eV}}{1.60 \times 10^{-19} \text{j}}$$
$$\sim 4 \times 10^{-7} \text{ eV}$$

We see from Fig. 1-4 that the natural breadths of the hydrogen energy levels are very much smaller than the energies themselves. When an excited state is long lived its natural width is even less significant. Long-lived states, called metastable, exist as long as 10^{-3} sec.

Additional broadening of spectral lines from a gas, emitting light, occurs for other reasons. Doppler broadening arises because of the thermal motion of the atoms. Collisional, or pressure, broadening occurs because energy levels in an atom are changed by the interaction during close encounters with other atoms.

1-10 THE ELECTROMAGNETIC SPECTRUM

Figure 1-10 shows the complete range of wavelengths and frequencies in the electromagnetic spectrum. Infrared, visible, and ultraviolet light result from changes in atomic and molecular configurations; gamma rays result from changes within the nucleus. Electromagnetic waves of longer wavelength (lower frequency) are generated by the slower oscillatory currents in antennas. Radar and radio waves are thus generated, and these emitted "carrier" waves can be modulated, for example, by relatively slow changing of the amplitude (AM) or frequency (FM) to transmit information.

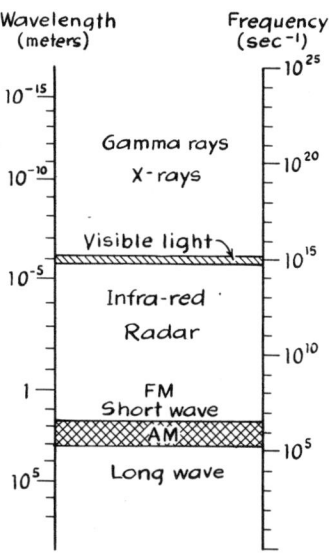

Fig. 1-10 Electromagnetic radiation spectrum.

PROBLEMS

1. Determine the atomic weight of chlorine from its natural composition: 75% $^{35}_{17}Cl$ and 25% $^{37}_{17}Cl$.
2. Calculate the average weight of a single atom of chlorine.
3. Why is the atomic weight of carbon given in Table 1-1 not exactly 12?
4. What is the "atomic weight" of an electron?
5. How many grams of sodium are needed to produce 5 grams of hydrogen in the reaction $2Na + 2H_2O \rightarrow H_2 + 2NaOH$? How many atoms of sodium enter the reaction?
6. How many pounds of uranium are contained in one ton of U_2O_3?
7. Calculate the number of atoms in one cm³ of boron (density = 2.45 g/cm³).
8. Using Avogadro's number and the atomic weight of lead (density = 11.3 g/cm³), calculate the volume occupied by a single lead atom.
[Ans. 3×10^{-23} cm³]
9. For a steady train of waves, having wavelength λ, and vibrational frequency ν, and moving with a velocity c, prove Eq. 1-1. (Hint. Imagine the passage of a steady stream of water waves past some reference point.)
10. In Eq. 1-2, show that Planck's constant h must have dimensions of energy × time.
11. Calculate the energy of a single X-ray photon of frequency 10^{19} vibrations per second.
[Ans. $6.6 \times 10^{-15} j$]

18 Electromagnetic Radiation and Atomic Structure

12. With the help of Eqs. 1-1 and 1-2 calculate the energy of a single photon of blue light ($\lambda = 5.40 \times 10^{-5}$ cm). [Ans. 3.7×10^{-19} j]
13. Calculate the wavelength of a gamma-ray photon of frequency 10^{24} vibrations per second. What is the photon energy?
14. Make a diagram of the atom $^4_2$He according to the Rutherford-Bohr model.
15. On what experimental observation was the Bohr model of the atom based? List Bohr's postulates.
16. Verify the numerical values given in Eqs. 1-6 to 1-8.
17. What is the total energy of the hydrogen atom with quantum number n, according to Bohr theory? Explain what is meant by the ionization continuum of an atom. What is the value of n at the edge of the continuum?
18. What is the ratio of the angular momentum of an electron in the $n = 3$ state of hydrogen to that in the ground state? What is the ratio of the electronic energies in these two states? [Ans. 3, 1/9]
19. What is the energy of the photon emitted by an electron when it makes a transition from the $n = 5$ state in hydrogen to the ground state? Calculate the wavelength of this light. In what region of the electromagnetic spectrum does it lie?
20. Calculate the energy needed to excite the electron in the hydrogen atom from its ground state to the state $n = 4$. Describe what physical change takes place in a hydrogen atom that absorbs a photon of this energy. [Ans. 12.75 eV]
21. Use Bohr's postulates as needed to derive an equation for the quantized radii for circular orbits when one electron orbits around two protons.
22. Use Eq. 1-7 to calculate the speed of the electron in the ground state of hydrogen. (When $n = 1$, $r = a_0 = 0.529 \times 10^{-10}$ m). What is the ratio of its centripetal acceleration (v^2/a_0) to the acceleration due to gravity ($g = 9.8$ m/sec^2)? (*Note.* This ratio gives a measure of the relative strength of the electrical forces in atoms and those of gravity).
23. Calculate the wavelength of a neutron having a velocity of 2200 m/sec. Is its wavelength larger at higher speeds? [Ans. 1.8 Å]
24. Verify Eq. 1-12 from Eq. 1-11.
25. Calculate the wavelength of a 5000 eV electron. At what energy is the wavelength of an electron 0.5 Å?
26. Use Eq. 1-10 to calculate the wavelength of a 10 gram object traveling with a speed of 4 m/sec.
27. Why is an X-ray photograph sometimes described as a "shadowgraph?"
28. Explain qualitatively from Eqs. 1-1 and 1-2 why higher voltage X-ray machines give lower values of λ_0 in Fig. 1-9.
29. Why are long wavelength X-rays not important in making an X-ray photograph? How can they be filtered out of the beam emerging from a machine?
30. Discuss the generation of X-rays in a television picture tube. (*Note.* X-rays are sometimes produced in greater quantity elsewhere in television receivers.)

31. Why are "lead aprons" sometimes worn by persons working in the vicinity of X-ray machines?
32. Discuss X-rays. Are they particles or waves? How do they relate to electronic structure of atoms? Draw a diagram of a Coolidge X-ray tube.
33. What is the value of λ_0 in Fig. 1-9 in an X-ray machine operating at 80,000 volts? [Ans. 0.15 Å]
34. By means of Eq. 1-14, estimate the minimum uncertainty in the momentum of an electron confined to an interval $\Delta x = 1$ cm. What is the uncertainty when the interval is $\Delta x = 1$ Å, the approximate diameter of the hydrogen atom in its ground state?
35. Estimate the natural line breadths of excited atomic states having lifetimes of 10^{-10} sec and 10^{-4} sec. [Ans. 4×10^{-5} eV, 4×10^{-11} eV]

CHAPTER TWO

Radiation and the Nucleus

The phenomenon of radioactivity is so varied, its properties are so diverse and so widespread in the universe that one must take it into consideration in studies of the natural sciences, in particular physiology and therapy, in meteorology and in geology.

Madame Curie

2-1 RADIOACTIVITY: ALPHA, BETA, AND GAMMA RADIATION

It will be recalled from Section 1-8 that the fluorescence of the cathode ray tube was under study when Roentgen first detected the presence of X-rays. Becquerel, in 1896, tried to discover whether X-rays are present around other sources of fluorescence. He used a number of fluorescent chemicals, one of which was a compound of uranium. After first exposing them to light, he wrapped these chemicals in black paper and placed them next to photographic film to see whether rays akin to those of Roentgen are generated.

Becquerel found that the films were darkened only when the uranium compound was used. To his surprise, he also noticed that the uranium salts blackened the film even when fluorescence had *not* been excited by first exposing the salt to light. The salts emitted some kind of radiation that left a silhouette of the crystals on the photographic film. Becquerel traced the source of the radiation to the element uranium itself. He found that the radiation continued to be emitted spontaneously from the element over long periods of time with apparently undiminished intensity. The phenomenon of spontaneous radiation from elements is called radioactivity, and materials showing this property are said to be radioactive. This radiation, like X-rays, also has the ability to ionize air and to excite fluorescent materials.

Following Becquerel's discovery, the Curies began a search for other radioactive materials. They discovered two previously unknown chemical elements, polonium and radium, by virtue of the intense radioactivity per unit mass of these elements. Radium is several million times more effective than uranium in blackening film. It is now known that all isotopes of elements with atomic numbers greater than that of bismuth (83) are radioactive. A few of the light elements have naturally occurring radioactive isotopes, or radioisotopes, for example, $^{3}_{1}H$, $^{14}_{6}C$, and $^{40}_{19}K$. In addition to these natural ones, about 600 man-made artificial radioisotopes have been produced. Some of the many uses of radioisotopes in industry, medicine, and biology will be described in later chapters.

Three common kinds of radiation are emitted from natural radioactive isotopes. As Fig. 2-1 shows, these can be distinguished both by their different paths in a magnetic field and by their ranges of travel in air. The rays emerge in a beam from a hole in a lead container and pass into a magnetic field, pointing perpendicularly into the plane of the figure as indicated by the circled crosses. Alpha particles, being positively charged,

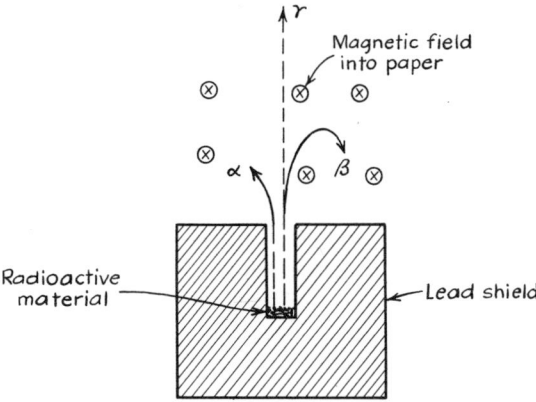

Fig. 2-1 Bending of radiation in a magnetic field. Alpha rays are doubly charged nuclei of the element helium, He^{++}, beta rays are electrons, and gamma rays are photons. Alpha, beta, and gamma rays originate in the nuclei of atoms.

are bent toward the left; and the negative beta rays, toward the right. Gamma rays pass undeflected through the magnetic field and hence are electrically neutral. The alpha rays are deflected much less than are the beta rays and travel only a few centimeters in air. Beta rays penetrate tens of centimeters; and gamma rays, several meters or more. The penetration of charged and uncharged radiation in matter is described in the next three chapters.

A fourth kind of radiation, a positive beta ray, or positron, is emitted by some artificial radioisotopes. All four types of radiation are emitted spontaneously from the nuclei of atoms, the particular kind or kinds and their energies depending on the radioisotope.

By collecting and neutralizing alpha rays, Rutherford found that helium atoms are formed. He showed that alpha rays are doubly charged nuclei of helium atoms, $_2^4He^{++}$. This finding is also confirmed by their observed charge-to-mass ratio. We shall denote the alpha particle by $_2^4\alpha$. Beta rays, which we symbolize by writing $_{-1}^0\beta$, are electrons ejected from the nucleus.[1] Positrons ($_1^0\beta$) are "positively charged electrons". Gamma rays ($_0^0\gamma$) are photons of electromagnetic radiation with properties described by Eqs. 1-1 and 1-2. Gamma rays are indistinguishable from X-rays, the difference between them relating only to their origin. Whereas X-rays originate outside the nucleus (Section 1-8), gamma rays are emitted from the nucleus itself.

Alpha, beta, and gamma rays have, like X-rays, the ability to ionize atoms, to cause fluorescence in certain materials and blackening in photographic film. All of these ionizing radiations can produce biological damage. In the past, visible burns occurred on the skin of workers with radioactive substances after prolonged contact with these materials. Ingested radioisotopes can also damage the body. Extensive evidence of internal damage in humans has been seen in workers in the radium watch dial painting industry. The common practice of tipping paint brushes on the tongue resulted in the ingestion of minute amounts of radium which subsequently metabolized into the bones. Studies and autopsies of radium dial patients have revealed a higher than normal incidence of certain malignances, such as bone sarcoma, that is correlated with the amount of radium found in the bone. Visible damage to the bone has occurred in individuals having as little as 10^{-6} g of Ra in the skeleton years after retirement from dial painting.

---

[1] Nuclear theory shows that electrons do not exist in the nucleus. Beta decay is described as a process in which an electron (or positron) is created and ejected from the nucleus.

2-2 NUCLEAR TRANSFORMATIONS ACCOMPANYING RADIOACTIVE DECAY

Emission of an alpha or beta particle changes the nuclear charge, or atomic number, of an atom. Such a transformation, therefore, changes an atom from one chemical element into another. The radioactive decay of an atom of the uranium isotope $^{238}_{92}U$, for example, which emits an alpha particle, is described by writing

$$^{238}_{92}U \rightarrow \, ^{4}_{2}\alpha + \, ^{234}_{90}Th \qquad (2\text{-}1)$$

The subscript and superscript totals balance on both sides of the arrow. The alpha particle, being a nucleus of helium ($^{4}_{2}He^{2+}$), removes a charge of +2 and a mass of 4 from the uranium nucleus, leaving an atom of atomic number 90 (thorium) and mass 234. As discussed more fully in Section 2-5, the decay of an atom by alpha particle emission into *two* definite products, as illustrated by Eq. 2-1 implies that the alpha-particle energy is unique. The spectrum of alpha particles from a radioisotope is limited to at most a few energies.

The thorium nucleus in Eq. 2-1 is radioactive and emits a beta particle and an antineutrino, which always accompanies beta decay:

$$^{234}_{90}Th \rightarrow \, ^{0}_{-1}\beta + \, ^{234}_{91}Pa + \, ^{0}_{0}\bar{\nu} \qquad (2\text{-}2)$$

The beta particle, being an electron, removes a charge of −1 and a negligible mass from the nucleus, leaving a nuclear charge of 91 (protoactinium). The antineutrino, denoted by $^{0}_{0}\bar{\nu}$, carries away energy but no charge and no detectable amount of mass. As discussed in Section 2-5, the "three-body" decay by beta emission, as illustrated by Eq. 2-2, leads to a continuous spectrum of beta-particle energies.

The protoactinium nucleus in Eq. 2-2 is also radioactive and decays to give

$$^{234}_{91}Pa \rightarrow \, ^{0}_{-1}\beta + \, ^{234}_{92}U + \, ^{0}_{0}\bar{\nu} \qquad (2\text{-}3)$$

The product of this decay is an atom of uranium with a mass number 234. The decay of $^{234}_{92}U$ and successive radioisotopes by alpha and beta particle emission ends finally in the production of the stable isotope $^{206}_{82}Pb$.

As already mentioned, the nuclei of some artificially produced radioisotopes emit positrons. These have the same mass as electrons and are frequently called positive electrons or positive beta particles. Positron emission, which is always accompanied by emission of a neutrino, decreases the atomic number of the nucleus by one unit and removes a negligible mass. Atoms of the arsenic isotope $^{72}_{33}$As, for example, decay into germanium by emission of a positron ($^{0}_{1}\beta$) and neutrino:

$$^{72}_{33}\text{As} \rightarrow {}^{0}_{1}\beta + {}^{72}_{32}\text{Ge} + {}^{0}_{0}\nu \tag{2-4}$$

The neutrino, $^{0}_{0}\nu$, like the antineutrino, takes away energy but no charge or mass. The existence of the neutrino and antineutrino was first postulated many years after the discovery of radioactivity, in order to account for the continuous nature of positive and negative beta-ray spectra. These particles interact extremely weakly with matter, and their direct experimental detection has been made only in recent years. Since negative electron emission is much more common than positive electron emission, we shall always imply the negative particle when referring to beta decay unless otherwise specified.

A few nuclei, classified as radioactive, capture an atomic electron from outside the nucleus, most often from the K-shell, and emit a neutrino. This results in the same atomic transformation as positron emission. Electron capture occurs in atoms of the palladium isotope $^{103}_{46}$Pd:

$$^{103}_{46}\text{Pd} + {}^{0}_{-1}\text{e} \rightarrow {}^{103}_{45}\text{Rh} + {}^{0}_{0}\nu \tag{2-5}$$

The rhodium atoms so formed emit a series of characteristic X-rays as electrons move closer to the nucleus to fill vacancies created by the captured *K*-shell electron.

Gamma rays do not change the atomic number or atomic mass number of the emitting nucleus. The emission of an alpha or beta particle often leaves the residual nucleus with an excess of energy, which it then emits in a short time (e.g., $\sim 10^{-13}$ sec) as one or more gamma photons. When gamma emission does not take place immediately, but later (e.g., after several minutes or hours), the atom is said to be metastable (Sec. 1-9). Like characteristic X-ray and optical spectra, which are different for different elements, the spectra of gamma rays emitted following radioactive decay are different for different nuclei. The analysis of gamma-ray spectra in a sample of material, therefore, permits the identification of the radioisotopes present. (See Fig. 2-4.)

2-3 RADIOACTIVE DECAY LAW

With the passage of time, fewer and fewer original atoms are left in a sample of a decaying radioisotope. As far as is known, radioactive atoms decay at random. In any short time interval Δt, the change ΔN in a large number of atoms of a radioisotope is proportional to the number of atoms N of the isotope and to the length of the time interval Δt:

$$\Delta N \propto -N\Delta t \tag{2-6}$$

The negative sign appears because N decreases, that is, the change ΔN in N is negative. Denoting the constant of proportionality by λ, which has the dimensions of reciprocal time (e.g., sec^{-1}) and is called the decay constant of the radioisotope, we write

$$\Delta N = -\lambda N\Delta t \tag{2-7}$$

This equation, describing a random process, is well known in mathematics. It implies that

$$N = N_0 e^{-\lambda t} \tag{2-8}$$

where N is the number of the original atoms of the radioisotope present at time t, N_0 is the number present at time $t = 0$, and $e = 2.718...$ is the base of natural logarithms. Equation 2-8 is called the exponential law of radioactive decay.[2] We shall see in the next chapter that a similar equation describes the random removal of photons from a beam of electromagnetic radiation as it penetrates matter.

The half-life τ of a radioisotope is the time taken for one-half of the number of atoms in a sample of the isotope to decay. Setting $N/N_0 = 0.5$ and $t = \tau$ in Eq. 2-8 gives

$$0.5 = e^{-\lambda \tau} \tag{2-9}$$

Taking the natural logarithm of both sides ($\ln 0.5 = -0.693$ and $\ln e^{-\lambda \tau} = -\lambda \tau \ln e = -\lambda \tau$), we find that

$$\tau = \frac{0.693}{\lambda} \tag{2-10}$$

In terms of half-life, Eq. 2-8 can be written

$$N = N_0 e^{-0.693 t/\tau} \tag{2-11}$$

[2] Equation 2-8 can be obtained from 2-7 with the help of calculus. Equation 2-7 is equivalent to the differential equation $dN/N = -\lambda dt$. Integrating, we obtain $\ln N = -\lambda t + C$ where ln denotes the natural logarithm and C is the constant of integration. Letting $N = N_0$ denote the number of atoms present at time $t = 0$, we have $\ln N_0 = C$. Thus we write $\ln N = -\lambda t + \ln N_0$, or $\ln (N/N_0) = -\lambda t$. It follows that $N/N_0 = e^{-\lambda t}$, and thus Eq. 2-8 is verified.

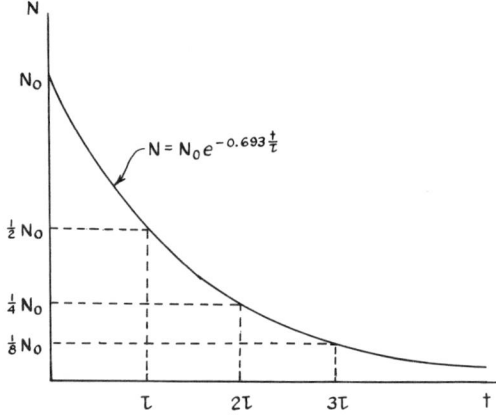

Fig. 2-2 Exponential law of radioactive decay. Starting at any time t, the curve N falls to one-half its value during the time interval between t and $t+\tau$.

Figure 2-2 shows a plot of N versus t as given by Eq. 2-11. After n half-lives the number of orginal atoms ($= N_0$ at $t = 0$) left in a sample of a radioisotope is

$$N = (\tfrac{1}{2})^n N_0 \tag{2-12}$$

The decay constant and half-life of a radioisotope can be determined by measuring the decrease in radioactivity of a sample with time. The rate of interaction of the emitted rays with a detector is proportional to the disintegration rate $-\Delta N/\Delta t$ in the sample which, in turn, is proportional to N (see Eq. 2-7). Thus the counting rate registered from a radioisotope decreases at exactly the same rate as N, and a plot of count rate versus time has the shape of the curve in Fig. 2-2 from which the half-life can be found.

2-4 NEUTRONS

Atomic nuclei are composed of positively charged protons and slightly heavier neutral particles called neutrons. The atomic number of an isotope is, by definition, the number of protons in the nucleus. Each proton and neutron contributes one unit to the atomic mass number of the isotope, and so the number of neutrons in a nucleus is equal to the difference in the atomic mass number and the atomic number. The nucleus of an atom of the fluorine isotope $^{19}_{9}F$, for example, consists of 9 protons and 10 neutrons.

The neutron was discovered by Chadwick in 1932. Shortly before this time it was known that bombarding the element beryllium with alpha rays produced a radiation more penetrating than gamma rays. This radiation liberates fast protons from hydrogeneous materials like paraffin, and the ionization from these could be readily detected. The lack of appreciable direct ionization implies that the beryllium radiation is electrically neutral, and it was thought for a time that it consisted of high-energy gamma photons. Another hypothesis was made that the radiation consists of a neutral particle, called the neutron, with a mass comparable to that of the proton.

By studying the recoil energies of protons and other nuclei bombarded with the radiation from beryllium, Chadwick showed that it did not consist of gamma photons. He determined the approximate atomic weight of the neutron from observation of the recoil nuclei struck by the radiation. The recoil particles, being charged, leave dense tracks of ions along their paths. Under conditions achieved in a cloud chamber, moisture from a supersaturated vapor condenses on the ions, and the tracks of the recoil nuclei can thus be made visible for quantitative observation (Fig. 2-3). The atomic weight of the neutron is 1.00867 as compared with the value 1.00728 for the proton. The nuclear reaction for producing neutrons ($_0^1n$) from bombardment of beryllium with alpha particles is written

$$_4^9\text{Be} + _2^4\alpha \rightarrow {}_6^{12}\text{C} + {}_0^1n \tag{2-13}$$

The total atomic and atomic mass numbers on both sides of the arrow balance. Chadwick neutron sources, made by mixing beryllium with an alpha emitter (e.g., radium or polonium), are still widely used.

Although neutrons produce almost no ions directly, we treat them as ionizing radiation because they produce ionizing recoil particles. Their passage through matter is described in Chapter 5. The recoil particles that they produce in tissue—for example, protons from hydrogen atoms—can cause biological damage to an organism. As we shall see in Chapter 8, neutrons appear to be one of the most effective kinds of radiation in producing biological damage.

The neutron is the key particle of the atomic age. Electrically neutral, it readily passes through the electronic cloud of an atom into the nucleus. The energy-releasing process of nuclear fission, for example, is triggered by a neutron, which enters a nucleus of high atomic number and causes it to split into smaller fragments. Three important fissionable isotopes, $^{233}_{92}\text{U}$, $^{235}_{92}\text{U}$, and $^{239}_{94}\text{Pu}$, produce several neutrons when they fission, and so it is possible to build a self-sustaining chain reactor. Thus neutrons are present in the environment of an operating reactor, and appropriate health protection measures for personnel are routinely taken. Particle

28 Radiation and the Nucleus

Fig. 2-3 Photograph of track of recoil proton from water vapor in cloud chamber irradiated by neutrons. A neutron produces almost no ions directly; its track, like that of an X- or gamma-ray photon, is not visible. Photograph from I. Curie and F. Joliot, Comptes Rendus, **194**, 876 (1932).

accelerating machines and cosmic rays also produce neutrons by means of other nuclear reactions.

Since neutron absorption does not change the atomic number of a nucleus, a large number of elements can be made radioactive, or "activated," in a reactor. Radiocobalt, ^{60}Co, a beta and gamma ray emitter having a half-life of 5.3 years, is made in the reaction

$$^{59}_{27}\text{Co} + ^{1}_{0}n \rightarrow ^{60}_{27}\text{Co} + ^{0}_{0}\gamma$$

when stable cobalt is put in a reactor. (Natural cobalt consists of 100%

^{59}Co). Entirely different elements can be produced as a result of neutron capture. Radiophosphorous, for example, is produced from sulfur ($^{32}_{16}$S, 95% abundant) by means of the reaction

$$^{32}_{16}S + ^{1}_{0}n \rightarrow ^{32}_{15}P + ^{1}_{1}H$$

in which a neutron knocks a proton ($^{1}_{1}$H) out of the nucleus. Neutron absorption does not necessarily produce a radioactive isotope. Stable $^{57}_{26}$Fe is formed when the nucleus of stable $^{56}_{26}$Fe (91.7% abundant) absorbs a neutron.

2-5 ENERGETICS OF NUCLEAR REACTIONS

Nuclear reactions result in an interchange of mass and energy in accordance with Einstein's formula

$$E = Mc^2. \tag{2-14}$$

Here M denotes the difference in mass before and after the reaction and E is the resulting change in energy. The quantity $c = 3 \times 10^8$ m/sec is the velocity of light in a vacuum. Energy is either released when mass is lost (exothermic reaction), or else it must be added to make the nuclear transformation possible (endothermic reaction). The amount of energy involved is called the Q-value of the reaction. Knowing isotopic masses, such as those listed in Table 2-1, we can, by means of Eq. 2-14, make quantitative predictions of energy changes that accompany alterations in the composition of atomic nuclei. Whereas the nuclei of heavy elements release energy when they break up, (e.g., in fission and in radioactive decay), those of light nuclei release energy when they combine (e.g., in the deuterium-tritium fusion reaction $^{2}_{1}H + ^{3}_{1}H \rightarrow ^{4}_{2}He + ^{1}_{0}n + 17.6$ MeV).[3]

Radioactive decay of an atom spontaneously releases a large amount of energy (compared, for example, with that released in a chemical reaction), which is shared among the daughter products and often partly transformed into gamma radiation. The total mass of the decay products is less than that of the parent atom, the mass difference being converted into energy. We now show how the energy released in different types of radioactive decay can be calculated from measured atomic masses.

---

[3]The unit 1 MeV = one million electron volts = 10^6 eV is commonly used in nuclear physics.

30 Radiation and the Nucleus

Table 2-1 Properties of Some Atoms and Atomic Particles

Particle or Atom	Atomic Mass (AMU) (Unified Scale)[a]	Half Life and Principal Radiation Emitted
electron	0.000549	stable
proton	1.00728	stable
neutron	1.00867	12.8 min (β)
^1_1H	1.00782	stable
^2_1H (deuterium)	2.01410	stable
^3_1H (tritium)	3.01605	12.26 yr (β)
^3_2He	3.01603	stable
^4_2He	4.00260	stable
^7_3Li	7.01601	stable
^9_4Be	9.01219	stable
$^{10}_5B$	10.01295	stable
$^{12}_6C$	12.000000...	stable
$^{14}_6C$	14.00323	5770 yr (β)
$^{14}_7N$	14.00308	stable
$^{222}_{86}Rn$	222.0165	3.82 day (α)
$^{226}_{88}Ra$	226.0244	1620 yr (α, γ)
$^{238}_{92}U$	238.0498	4.51×10^9 yr (α, γ)

[a] By definition, mass of $^{12}_6C$ atom is exactly 12 AMU.
1 Atomic mass unit (AMU) = 1.66×10^{-27} kg.
1 Atomic energy unit, electron volt: 1 eV = 1.60×10^{-19} j.
    1 million electron volts: 1 MeV = 1.60×10^{-13} j.
Equivalence of mass and energy ($E = Mc^2$): 1 AMU = 931 MeV.

Alpha Decay

As an example, we compute the mass lost and energy released (Q-value) in the decay of a radium nucleus by emission of an alpha particle. We perform the calculation first in MKS metric system units and then in atomic units.

Using the notation of Section 2-2, we write

$$^{226}_{88}Ra \rightarrow {}^4_2\alpha + {}^{222}_{86}Rn \tag{2-15}$$

By ejecting an alpha particle, which is a positively charged helium ion, $^4_2He^{++}$, the radium atom changes into a negatively charged radon ion, $^{222}_{86}Rn^{--}$. To show the presence or absence of atomic electrons, we write in place of Eq. 2-15.

$$^{226}_{88}Ra \rightarrow {}^{222}_{86}Rn^{--} + {}^4_2He^{++} \tag{2-16}$$

Energetics of Nuclear Reactions 31

The mass on the left-hand side of the arrow is the mass M_{Ra} of the neutral radium atom. The total mass on the right is the sum of (1) the mass M_{Rn} of the radon atom plus that of two electrons, $2m_0$, and (2) the mass M_{He} of the helium atom minus the mass of two electrons, $-2m_0$. The total mass on the right is thus $M_{Rn} + 2m_0 + M_{He} - 2m_0 = M_{Rn} + M_{He}$, the same as the total masses of the neutral radon and helium atoms. The mass difference M between the left and right sides of Eq. 2-16 is, therefore,

$$M = M_{Ra} - (M_{Rn} + M_{He})$$

In alpha decay the mass difference between parent atom and the daughter products is simply that of the neutral atoms.

Using Table 2-1, we find that, in atomic mass units,

$$M = 226.0254 - (222.0175 + 4.0026) = 0.0053$$
$$= 5.3 \times 10^{-3} \text{ AMU} \tag{2-17}$$

In the unified scale of atomic masses used in Table 2-1, 1 AMU = 1.66×10^{-27} kg and so, in MKS units,

$$M = (5.3 \times 10^{-3} \text{ AMU}) \times \left(1.66 \times 10^{-27} \frac{\text{kg}}{\text{AMU}}\right)$$
$$= 8.80 \times 10^{-30} \text{ kg}$$

From Eq. 2-14 it follows that the energy released in the decay is

$$Q = Mc^2 = (8.80 \times 10^{-30} \text{ kg}) \times \left(3 \times 10^8 \frac{\text{m}}{\text{sec}}\right)^2$$
$$= 7.92 \times 10^{-13} \frac{\text{kg-m}^2}{\text{sec}^2} = 7.92 \times 10^{-13} \text{ j.} \tag{2-18}$$

The energy Q is shared between the alpha particle, the recoiling radon ion, and any subsequent gamma rays from the radon nucleus. As shown by the decay scheme in Fig. 2-4, the nucleus of the isotope ^{226}Ra decays 94% of the time to the nuclear ground state of ^{222}Rn, with no subsequent gamma radiation. For these decays we may write

$$Q = \tfrac{1}{2}(M_{Rn} + 2m_0)V_{Rn}^2 + \tfrac{1}{2}(M_{He} - 2m_0)V_\alpha^2 \tag{2-19}$$

where V_{Rn} and V_α are the velocities of the recoiling radon ion and alpha particle. (In 5.7% of the decays an energy of 0.187 MeV is removed by a gamma photon, and this energy must be subtracted from the left-hand

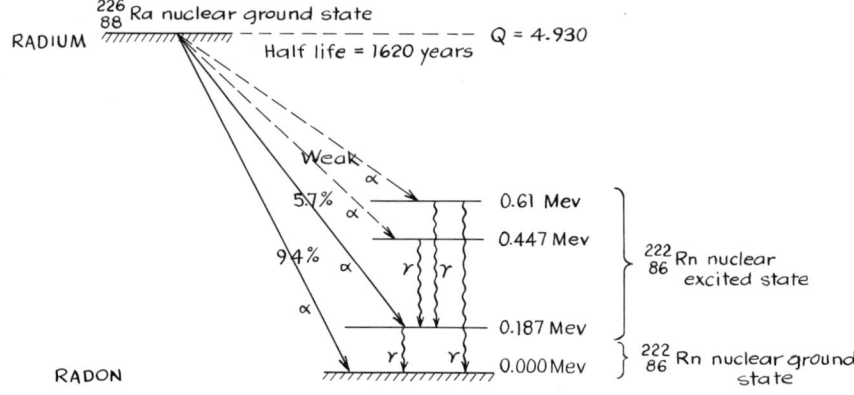

Fig. 2-4 Decay scheme, showing possible energy levels in which the $^{222}_{86}$Rn nucleus can be left as a result of alpha decay of the $^{226}_{88}$Ra nucleus. When alpha decay (slanted lines) leaves the radon nucleus in an excited state, it emits one or more gamma photons (vertical wavy lines). The presence of the isotopes $^{226}_{88}$Ra and $^{222}_{86}$Rn in a sample can be inferred from the energies and relative intensities of the emitted gamma radiation.

side of Eq. 2-19).[4] Because the radium atom is at rest before decay, the momenta of the products in Eq. 2-16 are equal and opposite (momentum conservation):

$$(M_{Rn} + 2m_0)V_{Rn} = (M_{He} - 2m_0)V_\alpha \tag{2-20}$$

Since the mass of two electrons is small compared with the total mass of an atom, we can, within a good approximation, neglect the presence of m_0 in Eqs. 2-19 and 2-20, writing

$$Q = \tfrac{1}{2}M_{Rn}V_{Rn}^2 + \tfrac{1}{2}M_{He}V_\alpha^2 \quad \text{and} \quad M_{Rn}V_{Rn} = M_{He}V_\alpha.$$

Substituting $V_{Rn} = (M_{He}/M_{Rn})V_\alpha$ from the second equation into the first, we find that

$$Q = \tfrac{1}{2}\frac{M_{He}^2}{M_{Rn}}V_\alpha^2 + \tfrac{1}{2}M_{He}V_\alpha^2 = \tfrac{1}{2}M_{He}V_\alpha^2 \frac{M_{He} + M_{Rn}}{M_{Rn}}$$

It follows that the kinetic energy of the alpha particle is given by

$$\tfrac{1}{2}M_{He}V_\alpha^2 = \frac{M_{Rn}}{M_{He} + M_{Rn}}Q = \frac{222}{226}Q = 0.982 Q$$

$$= 7.78 \times 10^{-13} \text{ j}. \tag{2-21}$$

[4]In contrast to the "three-body" decay by beta ray emission, in which a continuous spectrum of beta particle energies results (as we discussed following Eq. 2-2), an alpha particle from a radioisotope has only a few discrete energies.

With M_{He} expressed in kg we find that the speed of the alpha particle is

$$V_\alpha = \sqrt{\frac{2 \times 7.78 \times 10^{-13}}{M_{He}}} = \sqrt{\frac{2 \times 7.78 \times 10^{-13}}{4 \times 1.66 \times 10^{-27}}}$$

$$= 1.53 \times 10^7 \text{ m/sec}$$

Compared with the speed of light, $V_\alpha/c = 1.53 \times 10^7/3 \times 10^8 = 0.051$, and thus we are justified in using the classical, rather than relativistic, formulas for kinetic energy and momentum.

In atomic units we see from Eq. 2-17 and Table 2-1 that the energy released by the alpha decay of an atom of ^{226}Ra is $Q = (5.3 \times 10^{-3}\text{ AMU}) \times (931\text{ MeV/AMU}) = 4.93$ MeV. The kinetic energy of the alpha particle is, from Eq. 2-21, $0.982Q = 4.84$ MeV.

Beta Decay

We shall calculate the Q-value for the decay of the ^{14}C atom by emission of a beta particle from the nucleus:

$$^{14}_{6}\text{C} \rightarrow {}^{0}_{-1}\beta + {}^{14}_{7}\text{N} + {}^{0}_{0}\bar{\nu} \qquad (2\text{-}22)$$

As in Eq. 2-16 we indicate the number of atomic electrons associated with each member of Eq. 2-22 by writing

$$^{14}_{6}\text{C} \rightarrow {}^{0}_{-1}\beta^- + {}^{14}_{7}\text{N}^+ + {}^{0}_{0}\bar{\nu} \qquad (2\text{-}23)$$

The total mass on the right of the arrow is the sum of (1) the mass m_0 of the beta particle and (2) the mass M_N of the neutral nitrogen atom minus that of one electron, $-m_0$, the mass of the antineutrino being zero. The total mass on the right is thus $m_0 + M_N - m_0 = M_N$, the mass of the neutral nitrogen atom. As in alpha decay, therefore, the mass lost by the nucleus in beta decay is equal to the mass difference between the neutral parent and daughter atoms. The mass lost in Eq. 2-23 is, therefore, $M = M_C - M_N = 14.00323 - 14.00307 = 1.6 \times 10^{-4}$ AMU, where M_C denotes the mass of the atom $^{14}_{6}$C in Table 2-1. The Q value for the decay is $Q = 1.6 \times 10^{-4} \times 931 = 0.15$ MeV. This radioactive isotope of carbon, which has a half-life of 5770 years and is produced by bombardment of atmosphere nitrogen by neutrons according to the reaction

$$^{14}_{7}\text{N} + {}^{1}_{0}n \rightarrow {}^{14}_{6}\text{C} + {}^{1}_{1}\text{H}, \qquad (2\text{-}24)$$

is used in radiocarbon dating (see Problem 11).

When no gamma radiation is emitted from the nucleus after beta decay, the energy Q is shared among three decay products: the beta ray, the antineutrino, and the daughter nucleus. The spectrum of beta ray energies is continuous, ranging from zero to the Q-value. The maximum energy of a beta ray from $^{14}_{6}C$ is thus 0.15 MeV. For many isotopes, the average energy of a beta particle is $\sim Q/3$.

Positron Decay

To show the number of atomic electrons in Eq. 2-4, we write

$$^{72}_{33}As \rightarrow {}^{0}_{1}\beta^{+} + {}^{72}_{32}Ge^{-} + {}^{0}_{0}\nu \tag{2-25}$$

The total mass on the right of the arrow is the sum of (1) the positron mass, m_0, and (2) that of the germanium ion, which is $M_{Ge} + m_0$, M_{Ge} being the mass of the neutral ^{72}Ge atom. The total mass on the right in Eq. 2-25 is $m_0 + M_{Ge} + m_0 = M_{Ge} + 2m_0$; the mass lost in the decay, therefore, is given by

$$M = M_{As} - (M_{Ge} + 2m_0) = M_{As} - M_{Ge} - 2m_0 \tag{2-26}$$

where M_{As} denotes the mass of the ^{72}As atom. Unlike alpha and beta decay, the mass lost in positron emission is *not* simply the mass difference of the neutral parent and daughter atoms. As Eq. 2-26 shows, twice the electron mass must be subtracted from this difference in calculating the Q-value.

Because a neutrino is always emitted with a positron, positron energy spectra, like those of beta particles, are continuous, ranging from zero to the Q-value for the decay.

PROBLEMS

1. List three methods for detecting each of the radiations shown in Fig. 2-1.
2. Write an expression like Eq. 2-1 for the decay of $^{222}_{86}Rn$ into polonium (Po) by emission of an alpha particle.
3. Write an expression like Eq. 2-2 for the beta decay of tritium (^{3}H).
4. As stated in the text after Eq. 2-3, a series of radioisotopes begins with $^{238}_{92}U$ and ends with stable $^{206}_{82}Pb$. How many alpha and beta particles are emitted in the transformation of one atom of $^{238}_{92}U$ into one of $^{206}_{82}Pb$?
5. Why must the number of atoms N in Eq. 2-6 be large?
6. Calculate the number of atoms that decay per second in a sample of a radioisotope that contains 4×10^{20} atoms and has a decay constant of 6×10^{-12} sec^{-1}.

Problems 35

7. A sample of $^{222}_{86}$Rn (half-life = 3.82 days) contains 10^{16} atoms at time $t = 0$. How many atoms of ^{222}Rn will be left after 19.1 days? [Ans. 3.12×10^{14}]
8. The half-life of $^{226}_{88}$Ra is 1680 years. What is the decay constant of this isotope in sec^{-1}?
9. A pure sample of radioactive material contains 2.045×10^{23} atoms. If 10^6 atoms decay per second, determine the half-life of the material in years. [Ans. 4.5×10^9 yrs.]
10. Prove Eq. 2-12.
11. The half-life of the radioactive element ^{14}C if 5770 years. If a sample were found to have 10^{16} atoms of ^{14}C in 1967, how many atoms of ^{14}C were present in the year 9573 BC? How many atoms decayed per month in the year 9573 BC? [Ans. $4 \times 10^{16}, 4 \times 10^{11}$]
12. Explain how a radioactive isotope (such as $^{226}_{88}$Ra) can be used to make a luminous watch dial.
13. How many protons and how many neutrons are there in the nucleus of an atom of $^{23}_{11}$Na?
14. A nucleus of the stable nuclide $^{10}_{5}$B absorbs a neutron and breaks up into an alpha particle and a lithium (Li) nucleus. Write this reaction in symbols.
15. Write the reaction showing how radioactive iron ^{55}Fe is produced from stable ^{54}Fe (5.8% abundant) by neutron absorption.
16. How much energy is released when one gram of matter is converted into energy? [Ans. 9×10^{13} j]
17. By how much does the mass of 1 g of water increase when it is heated 1°C?
18. The fission of one nucleus of $^{235}_{92}$U releases about 200 MeV of energy. Estimate the number of fissions occurring per second in a nuclear reactor that operates at a power of 30 megawatts (3×10^7 watts = 3×10^7 j/sec). [Ans. 10^{18} sec^{-1}]
19. The energy released in the radioactive decay

$$^{238}_{92}\text{U} \rightarrow {}^{4}_{2}\alpha + {}^{234}_{90}\text{Th}$$

is 4.27 MeV. Calculate the mass of ^{234}Th.
20. In the radioactive decay of a certain nucleus, 5.4×10^{-3} atomic mass units are converted into energy. What is the Q value? Assume that the momentum of the nucleus before decay is zero, and that the decay products are particles of mass M and $2M$. What are the kinetic energies of the particles? [Ans. 5.0 MeV, $2Q/3$, $Q/3$]
21. What fraction of the energy released in Problem 19 is transformed into kinetic energy of the thorium atom as it recoils in the decay? How much energy does the alpha particle receive when the decay leaves the thorium nucleus in its ground state? [Ans. 0.0168, 4.2 MeV]
22. In the slow neutron reaction,

$$^{1}_{0}n + {}^{10}_{5}\text{B} \rightarrow {}^{7}_{3}\text{Li} + {}^{4}_{2}\text{He} + Q \quad (Q = 2.8 \text{ MeV}),$$

calculate the kinetic energy of $^{4}_{2}$He. [Ans. 1.8 MeV]

36 Radiation and the Nucleus

23. What is the maximum energy of a beta particle released in the decay of tritium, $^3_1\text{H} \rightarrow {}^{\,0}_{-1}\beta + {}^3_2\text{He} + {}^0_0\bar{\nu}$?
24. What is the Q-value for the spontaneous decay of a free neutron, $^1_0n \rightarrow {}^{\,0}_{-1}\beta + {}^1_1\text{H}^+ + {}^0_0\bar{\nu}$? [Ans. 0.78 MeV]
25. Calculate the Q-value of the reaction $^4_2\alpha + {}^9_4\text{Be} \rightarrow {}^1_0n + {}^{12}_6\text{C}$. Is the reaction exothermic or endothermic?
26. A radioactive sample undergoes alpha decay with particles having energy of 6 MeV. How much energy (MeV) is emitted by the source in two days if it contains 10^{18} atoms at the beginning and has a half-life of one day. [Ans. 4.5×10^{18} MeV]

CHAPTER THREE

Interactions of Charged Particles with Matter

There seems to be no doubt that such swiftly moving particles pass through the atoms in their path, and that the deflexions observed are due to the strong electric field traversed within the atomic system.

E. Rutherford

3-1 INTRODUCTION

Charged and neutral particles and electromagnetic waves penetrate matter and alter it in various ways. In this chapter and the next two chapters we discuss the ways in which these radiations interact with matter and the physical effects these interactions have. These effects are important both because they give insight into chemical and biological actions of radiation, and because they provide a basis for the detection and measurement of radiation.

3-2 EXAMPLES OF CHARGED PARTICLES

Protons occur as cosmic rays coming from outer space and may be very energetic. Some cosmic ray protons have energies in excess of 1 GeV, (i.e., 10^9 eV), and therefore can travel large distances in matter before coming to rest. For example, a 200 MeV proton can penetrate about 25 cm of water or about 12 cm of aluminum. Protons can be produced in machines from ionized hydrogen and accelerated up to energies of many GeV.

Alpha particles occur in cosmic rays and are produced in accelerators in the manner described above for protons. In addition, alpha particles occur in nature in another important way: they occur as the result of *radioactive decay* of unstable atoms. Some examples of radioactive decay were given in Chapter 2. When they occur in radioactive decay they have energies ranging from 4 to 10 MeV. Since alpha particles can come from about 160 different nuclides, they always present a potential source of radiation which requires that special health protection procedures be employed. In the energy range between 4 and 10 MeV, alpha particles are not very penetrating and cannot pass through thin layers of metal foil or even the human skin.

When radioactive elements undergo certain kinds of decay, electrons are emitted; these electrons are called *beta rays* and may have a considerable spread of energy depending on the elements from which they come. Beta rays from a given element, or emitter, do not all have the same energy; in fact they have energies ranging all the way from the maximum value, characteristic of the given emitter, to zero energy. High energy electrons also originate from cosmic ray interactions with the earth's atmosphere and in man-made accelerators, and have exactly the same properties as do beta rays. Electrons are more penetrating than either protons or alpha particles of the same energy. A 200 MeV electron comes to rest after penetrating about 50 cm of water or about 17 cm of aluminum.

3-3 EXCITATION AND IONIZATION OF MATTER

According to atomic theory each electron in the atom is in a discrete, or quantized, energy state (see Fig. 3-1). If the electrons fill levels at which their total energy is at its minimum value, the atom is said to be in its *ground state*. All atoms have unfilled levels to which electrons can be "promoted" by energy absorption. When electron promotion occurs, the atom is said to be *electronically excited*. If enough energy is absorbed to remove an electron from the influence of the nucleus, the atom may be *ionized*. The minimum energies which electrons in various levels must acquire to be freed from the atom are called their *ionization potentials*. When an electron absorbs energy in excess of its ionization potential, this excess appears as the electron's kinetic energy. Free electron energies are not quantized; any positive value is possible. This continuous region of energies, which is called the ionization continuum, is shown in Fig. 3-1.

When an alpha, beta or other energetic charged particle approaches an atom or molecule several processes of energy loss occur. Fundamentally, the interaction leading to energy loss results from the electric forces between the charged particle and the electrons in the atoms. Energy losses are due essentially to electronic excitation and ionization of atoms. When atoms are joined to form molecules, the passage of a charged particle can cause the atoms to vibrate at an increased rate or even to come apart, (i.e., dissociate). Dissociation of the molecule is frequently a process which competes with ionization. When charged particles pass

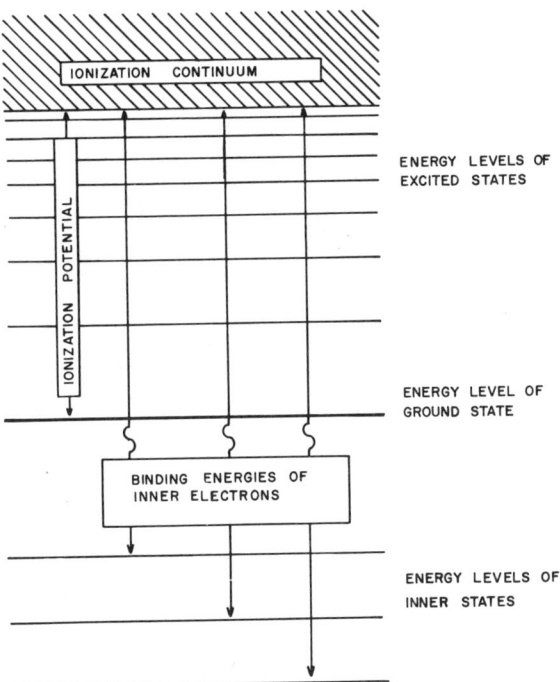

Fig. 3-1 Schematic energy level diagram of an atom. Ionization potential of valence electron varies from about 4 to 20 eV depending on atom. Binding energies of inner electrons are frequently several thousand electron volts.

very close to the nucleus of an atom, nuclear collisions or reactions may result. However, the most important energy loss processes are the electronic losses (excitation and ionization) discussed above. In fact, the valence, or most weakly bound, electrons account for most of the energy lost by an energetic charged particle. Occasionally, however, an inner electron is removed. When this happens, other electrons fill the "hole" left by the removed inner electron. The process in which atoms readjust to the removal of an inner electron causes what is called an *Auger cascade*, since the net effect may be the removal of several electrons from the same atom. Auger cascades in molecules have been observed in which all electrons are stripped from the molecule. When this happens, for example, if all of the 10 electrons are removed from CH_4, the positively charged nuclei fly apart in a process referred to as a Coulombic explosion.

3-4 STOPPING POWER OF CHARGED PARTICLES IN MATTER

Charged particles of the same energy and charge, but of different masses, penetrate different distances in matter, the lighter particles traveling farther. To discuss this effect in a more quantitative fashion we define *stopping power, dE/dX*, of a charged particle, as the average amount of energy lost by the particle per unit distance along its direction of travel. Since energy loss events occur at random the average is taken over a large number of particles at a given energy.

Let us imagine a charged particle passing through matter which we represent as a collection of atoms. Figure 3-2 illustrates the interaction of a charged particle with a single atom. The charged particle will exert a force on the electrons in the atom which is electrical in nature, that is by virtue of the charge on the particle it interacts with the electrons of negative charge according to Coulomb's law. The force (or strength of the

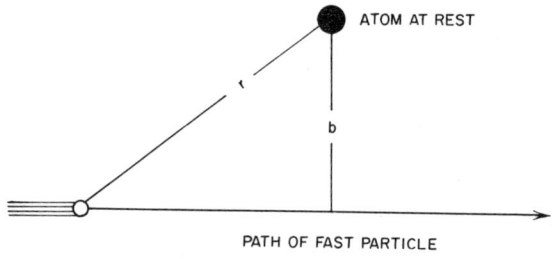

Fig. 3-2 Interaction of a charged particle with an atom. At arbitrary time the atom to particle separation is r. Distance of closest approach is b and is called the impact parameter.

interaction) depends on the charge of the moving particle, the number and distribution of electrons in the atom, and the separation of atom and particle. The energy lost by the particle will depend on the *force as a function of distance*, the impact parameter, and the velocity of the charged particle. Thus, the ability of the particle to lose energy to the medium depends on the type of matter, the charge of the moving particle, its velocity, and the impact parameter. The calculation of the stopping power, denoted by dE/dX, is beyond the scope of this book, but we can understand its general properties in a qualitative fashion. The quantity dE/dX is found by averaging energy losses over all possible values of the impact parameter, thus leaving a dependence of stopping power on the medium, the charge of the particle and its velocity. It is customary to specify dE/dX as a function of the particle energy, rather than its velocity.

Figures 3-3 and 3-4 give values of the stopping power of electrons and protons in H_2O, and for comparison in two metals, Al and Pb, for various values of the particle's energy. It is seen that, for a given energy (e.g., 1 MeV), the stopping power is lower for electrons than for protons. Basically the reason for this effect is that electrons pass a given atom in the medium much more quickly than do protons of the same energy, and because it interacts with the atomic electrons for a shorter time, its energy loss is smaller. Similar arguments explain why the stopping power is less for protons than for the heavier alpha particles. An examina-

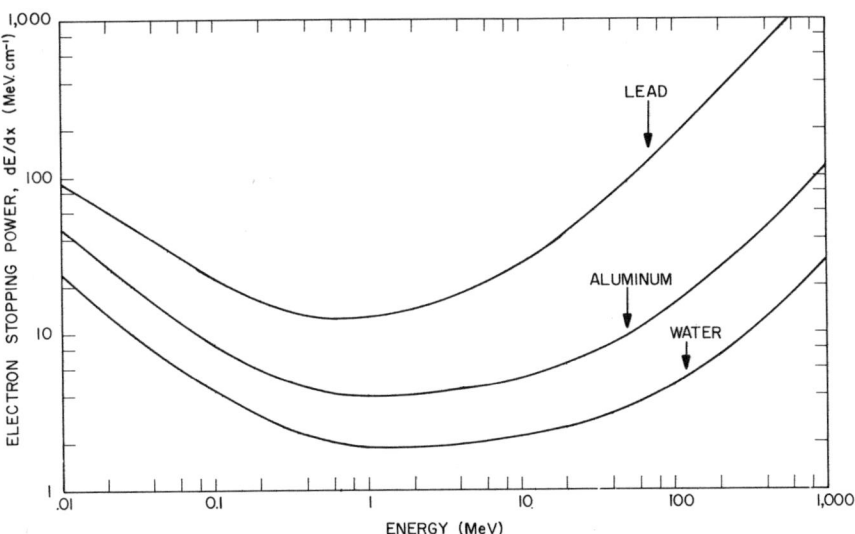

Fig. 3-3 Stopping power of water, aluminum (density, 2.7 g cm^{-3}) and lead (density 11.3 g cm^{-3}) for electrons as a function of energy.

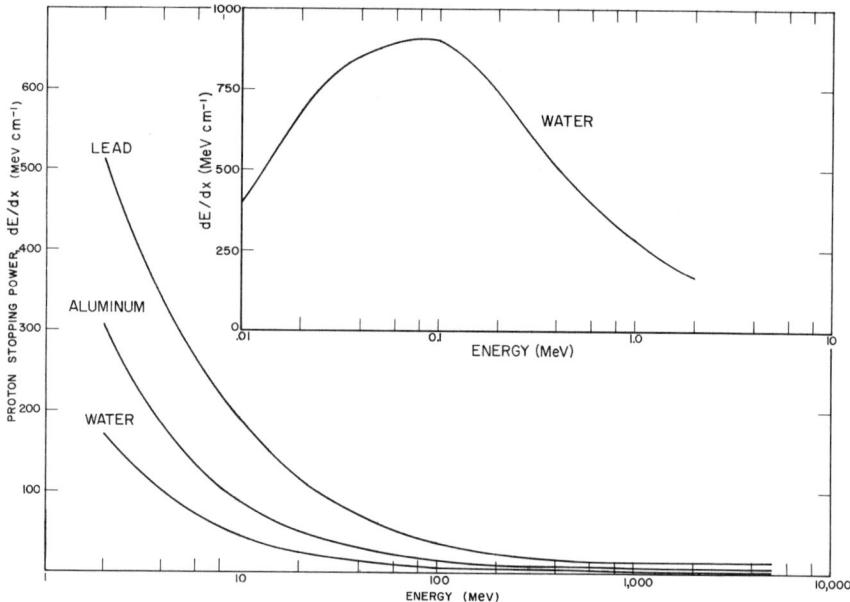

Fig. 3-4 Stopping power of water, aluminum (density, 2.7 g cm^{-3}) and lead (density 11.3 g cm^{-3}) for protons as a function of energy.

tion of Figures 3-3 and 3-4 also illustrates the point made above that dE/dX depends on the kind of medium. The more dense media, containing more electrons per unit volume, have a higher stopping power. Further examination of Fig. 3-3 or 3-4 shows that dE/dX generally increases as the velocity (or energy) of the given particle decreases, and this again is in agreement with our qualitative discussion. However, dE/dX reaches a maximum value (Bragg peak) and begins to decrease as the particle energy is further decreased, see Fig. 3-4. This effect is caused, in part, by the fact that positively charged particles capture electrons when the velocity of the particle is low. Thus the charged particle tends to become less charged and interact less strongly with the bound electrons. Another factor decreasing dE/dX at low velocities is that the electrons have time to readjust their positions in the atoms in such manner that electronic excitation is less likely.

3-5 RANGE OF CHARGED PARTICLES IN MATTER

The range of charged particles at a given energy is defined as the average distance they travel before they come to rest. The range is small when dE/dX is large, and conversely the range will be large when dE/dX

is small. The range of a particle whose energy is E_0 can be calculated when dE/dX is known for all values of energy between 0 and E_0. As stated above, energy loss events are statistical, hence particles with the same energy will travel slightly different distances before coming to rest, a phenomenon called range straggling.

Figures 3-5 and 3-6 give the range of electrons and protons in H_2O, Al, and Pb for several values of the particle's initial energy. Just as we expect from Figs. 3-3 and 3-4 the range of electrons is much greater than the range of protons for a given initial energy and for a specific medium. Similarly, a particle of even shorter range is the alpha particle, Fig. 3-7. The range of all of the particles is greatest in H_2O, intermediate in Al, and smallest in Pb.

Data on ranges of charged particles are of practical importance. Even at low energies electrons can penetrate deeply into living tissue (see the data for water). Protons are less penetrating than electrons; however,

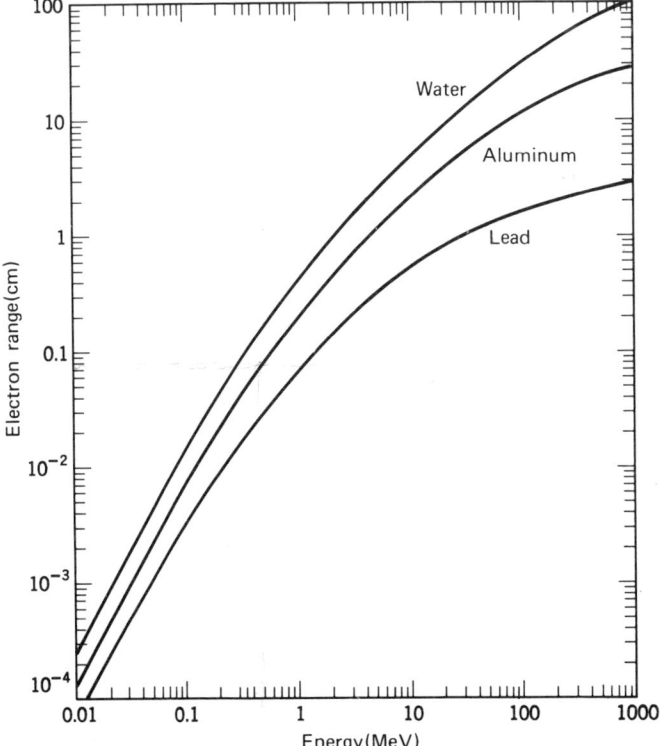

Fig. 3-5 Range of electrons in water, aluminum (density, 2.7 g cm^{-3}) and lead (density 11.3 g cm^{-3}) as a function of energy.

44 Interactions of Charged Particles with Matter

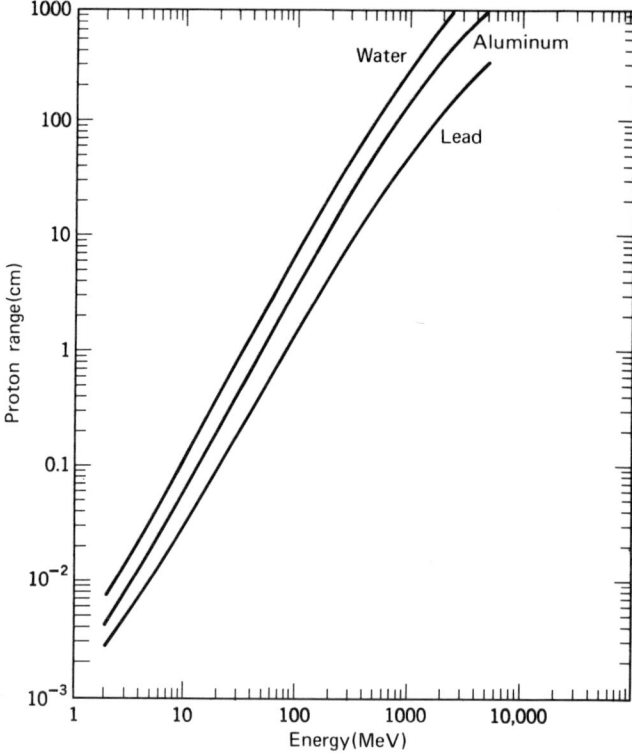

Fig. 3-6 Range of protons in water, aluminum (density, 2.7 g cm^{-3}) and lead (density, 11.3 g cm^{-3}) as a function of energy.

when the energy exceeds about 200 MeV, protons can penetrate a distance in soft tissue equivalent to the thickness of an average man. Alpha particles do not penetrate the skin, except when their energy is greater than 5 MeV. The data in Figs 3-5, 3-6, and 3-7 also show that heavy materials are efficient shields for charged particles.

PROBLEMS

1. By some means 10 eV of energy is given to a valence electron which is bound to an atom by 8 eV. What is the energy of the electron in the ionization continuum? If the atom has an excited state at 5 eV, how much energy is required to eject an electron that can just excite another identical atom in its ground state? [*Ans.* 2 eV, 13 eV]

2. A 2000 Å photon is emitted from an atom in returning to its ground state. What was the energy level of the excited state?

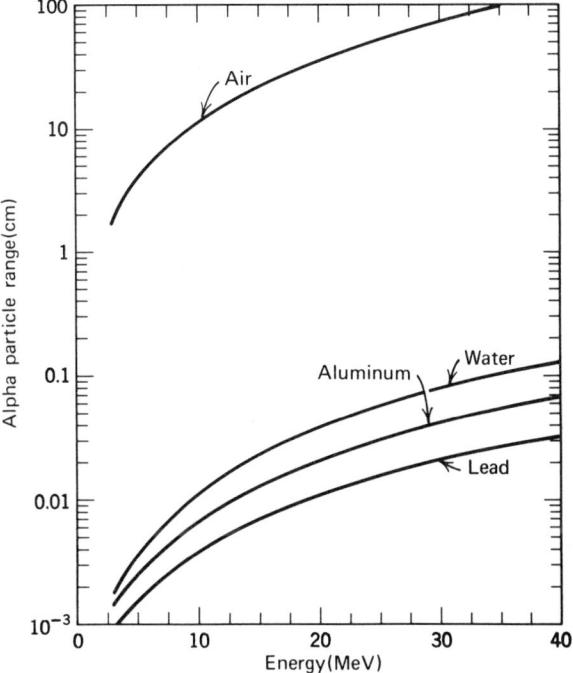

Fig. 3-7 Range of alpha particles in air (at 760 torr and 15°C), water, aluminum (density, 2.70 g cm^{-3}) and lead (density, 11.3 g cm^{-3}) as a function of energy.

3. The stopping power of lead for an electron is 22 MeV cm^{-1}. If the electron's energy were increased 60 times, the stopping power would also be 22 MeV cm^{-1}. What is the electron's energy? [*Ans.* 0.1 MeV]

4. What are the possible values of the stopping power of aluminum and water for an electron whose stopping power in lead is 40 MeV cm^{-1}?

5. The stopping power of aluminum for protons is 200 MeV cm^{-1}. What is the proton's energy? For this energy determine the values of dE/dX in lead and in water in units of MeV g^{-1}cm^2.
  [*Ans.* 3.5 MeV, 33 Mev g^{-1} cm^2, 112 Mev g^{-1} cm^2]

6. If the range of an electron in lead is 1 mm, what is its range in aluminum and in water? Assume that a maximum of 1 cm of material can be used as a shield against electrons. If it is required that no electrons be transmitted, what would be the maximum permitted energy of the electrons, when the material is, in turn, water, aluminum, and lead?
  [*Ans.* 2.1 mm, 4.2 mm, 32 MeV, 48 MeV, 78 MeV]

7. Repeat Problem 6 for protons instead of electrons.

8. At 10 MeV the stopping power of water, aluminum, and lead for protons is 50 MeV/cm, 100 MeV/cm, and 200 MeV/cm respectively. Estimate the

energy losses in samples of each of the materials for a thickness of 10^{-2} cm. Which estimate is the most accurate? In which material would the range of the 10 MeV proton be the greatest?

9. Calculate the pressure of air at 15°C at which 5 MeV alpha particles have a range of 10 cm. [*Ans.* 266 torr]

10. What is the range of 10 MeV alpha particles in tissue, assuming that water and tissue are equivalent? What thickness of aluminum is required to completely absorb 5 MeV alpha particles?

11. List three important factors that determine the stopping power of a charged particle.

12. Describe an experiment in which thin foils and a proton detector could be used to determine the energy of protons emerging from an accelerator.

13. Describe what would happen following the removal of two electrons from the molecule H_2. Explain why electrons may be called nature's glue.

CHAPTER FOUR

Interactions of X-Rays and Gamma Rays with Matter

A piece of sheet aluminum, 15 mm thick, still allowed the X-rays (as I will call the rays, for the sake of brevity) to pass, but greatly reduced the fluorescence. Glass plates of similar thickness behave similarly; lead glass is, however, much more opaque than glass free from lead. Ebonite several centimeters thick is transparent. If the hand be held before the fluorescent screen, the shadow shows the bones darkly, with only faint outlines of the surrounding tissues.

W. C. Roentgen

4-1 INTRODUCTION

The penetration of *electromagnetic radiation* (e.g., X-rays or gamma rays) into matter is somewhat more complex than the considerations given above for *charged particles*. We saw above that a charged particle — by virtue of the electrical force which it can exert on electrons in a molecule — can excite, ionize, or dissociate a molecule. Since electromagnetic radiation is not charged one might suppose that it would pass through

48 Interaction of X-Rays and Gamma Rays with Matter

matter with little interaction. A photon of the same energy as the charged particles considered (electrons, protons, and alpha particles) *does* penetrate further than even the electron. However, it is well known that X-rays and gamma rays are *diminished in number* on passing through matter, particularly through materials of high atomic number. Also, it is known that some of the X-rays or gamma rays *may* emerge from an absorbing medium with their *initial energy*, though the number which penetrated decreases with the thickness of the absorbing material. Clearly, then, our consideration of photons in matter must be based on some additional physical processes. In our range of interest, that is, photons having energies greater than a few keV, physical processes of three types are of greatest importance; the photoelectric effect. Compton scattering, and pair production.

4-2 CROSS SECTION

When discussing the probability, or chance, that electromagnetic radiation or uncharged particles will interact on passing through a specified thickness of matter, it is very convenient to make use of a concept known as "cross section". Therefore, let us suppose that in Fig. 4-1 particles (or photons) are parallel and are normally incident to a slab of material of a certain thickness. If the slab is thin enough, increasing its thickness will increase the chance that a particle will interact with it in like proportion. A slab of materia meeting this requirement is known as a "thin" slab.

Now, let us assume that the slab in Fig. 4-1 is thin and has thickness ΔX. When n_0 particles are normally incident to the slab n' "interactions" will occur. Evidently,

$$n' \propto n_0 \Delta X \tag{4-1}$$

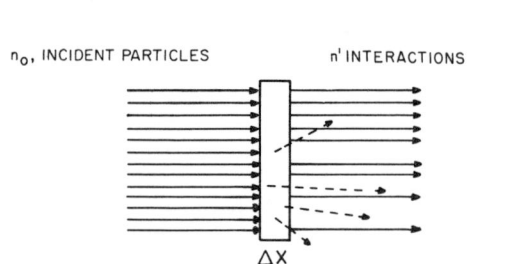

Fig. 4-1 Definition of cross section. When n_0 particles are incident on a material of ρ_A atoms per cm³ and of thickness ΔX the number of particle interactions is given by $n' = n_0 \sigma \rho_A \Delta X$ where σ is the cross section.

since n' must be proportional to both n_0 and ΔX. Equation 4-1 does not contain the number of atoms in the material, thus to introduce the known atomicity of matter we may write

$$n' = \sigma n_0 \rho_A \Delta X \qquad (4\text{-}2)$$

where ρ_A is the number of atoms per cm³ of material, and σ is a constant of proportionality. In Eq. 4-2, σ must have dimensions of cm² (ΔX has dimensions of cm), and for this reason σ is called cross section. Thus, the cross section of an atom is the area which the atom presents to the incoming particle for an interaction of a specified type.

Cross sections are not atomic constants, for they also depend on the type of incident particle, the energy of the particle, and the type of interaction. In some cases cross sections are calculated from theory, but in most cases they are determined from experimental data. Figure 4-1 and Eq. 4-2 suggest how the experiments are done; n_0 particles are directed at a thin slab of known ΔX and ρ_A. Measurement of the number of interactions of particular interest, n', is sufficient for calculation of σ from Eq. 4-2.

4-3 PHOTOELECTRIC EFFECT

Cross sections for the photoelectric effect in aluminum ($Z = 13$), iron ($Z = 26$), tellurium ($Z = 52$), and lead ($Z = 82$), as a function of photon energy is shown in Fig. 4-2. Consistent with Eq. 4-2, the units of cross section are cm² per atom except for the factor 10^{-24}. The latter factor is almost always used in nuclear physics and the magnitude 10^{-24} cm² is called a "barn", thus the scale in Fig. 4-2 ranges from 1 to 10^5 barns The photoelectric effect cannot occur between a photon and a free electron, that is, the electron must be bound in an atom. Furthermore, the more complex the atoms, that is, the greater the atomic number, the more likely it is that the photoelectric effect will occur.

In the photoelectric effect a photon gives up all its energy to an electron bound in an atom. Thus, the kinetic energy, T, of the ejected electron is given by

$$T = h\nu - B \qquad (4\text{-}3)$$

where $h\nu$ is the energy of the incoming photon and B is the electron binding energy, or ionization potential of the ejected electron. In many cases the energy B is much less than the energy $h\nu$ and the "photoelectron" carries off nearly all of the energy of the incoming photon. We are

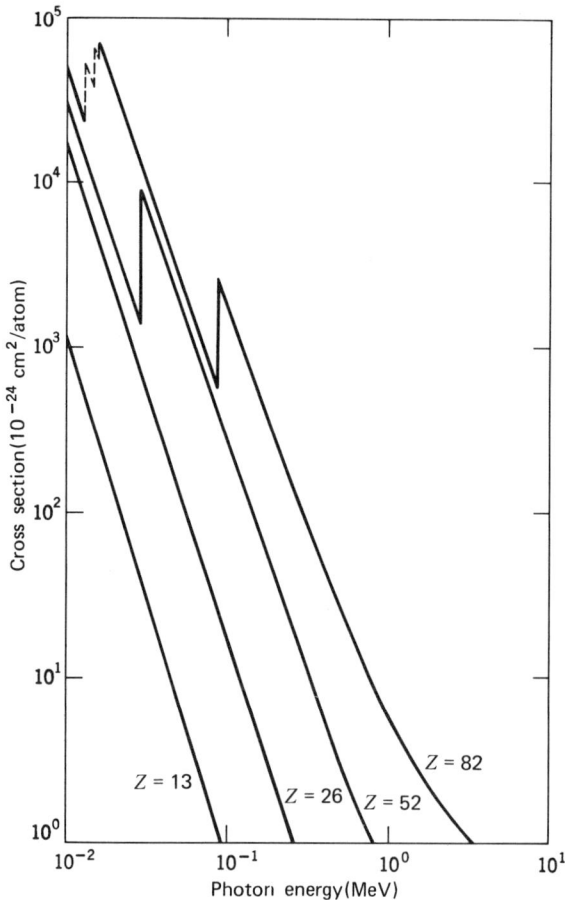

Fig. 4-2 Cross sections for the photoelectric effect, in various materials (having Atomic Number Z) as a function of photon energy. (From K. Z. Morgan and J. E. Turner, eds., *Principles of Radiation Protection*, John Wiley and Sons, 1967).

left then with an electron which further interacts with matter; our discussion above on the ranges of electrons and beta rays applies equally well to "photoelectrons."

4-4 COMPTON EFFECT

Unlike the photoelectric effect, the Compton effect can occur between a photon and a free electron. In the Compton effect the photon loses only a fraction of its energy. The energy that is lost is imparted to a free electron. After each interaction a photon and an electron are scattered at angles with respect to the original photon direction. The energy imparted to the free electron depends on the angle of scatter and the incoming photon energy; typically the scattered electron carries off one-half of the incident photon energy.

The cross section for Compton scattering for free electrons is shown in Fig. 4-3. Note should be made of the fact that the cross section is stated as cm²/electron, which emphasizes that the Compton effect is nearly independent of the binding of an electron in the atom. To find the total Compton cross section for an atom containing Z electrons the values in Fig. 4-3 are multiplied by the atomic number, Z. As Z becomes large the photoelectric effect will dominate over the Compton effect at moderate energies.

4-5 PAIR PRODUCTION

Compton scattering and the photoelectric effect are both examples of conversion of photon energy into mainly kinetic energy of electrons. The third important effect, *pair production*, involves also the conversion of photon energy to the *creation of matter*. In our range of interest, photons can create the lightest particle, the electron, and its antiparticle the

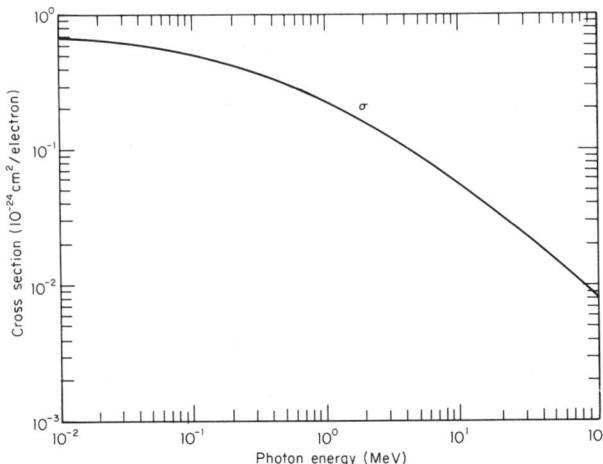

Fig. 4-3 Cross section for the Compton process as a function of photon energy. (Based on K. Z. Morgan and J. E. Turner, eds., *Principles of Radiation Protection*, John Wiley and Sons, 1967).

positron, both of which have a rest mass m_0. The Einstein equation states that the minimum photon energy is

$$E_{min} = 2m_0c^2 \qquad (4\text{-}4)$$

where c is the speed of light. The energy conservation equation for the pair production process is at energy $h\nu > E_{min}$

$$h\nu = 2m_0c^2 + T^+ + T^- \qquad (4\text{-}5)$$

where T^+ and T^- are the kinetic energies of the positron and electron, respectively. The value of $2m_0c^2$ is 1.02 MeV, thus pair production does not occur unless the photon energy is greater than 1.02 MeV. When the photon energy is greater than 1.02 MeV pair production *can* occur and the kinetic energy of the electrons and positrons are determined from Eq. 4-5. The cross section for pair production increases with atomic number Z and with photon energy as shown in Fig. 4-4.

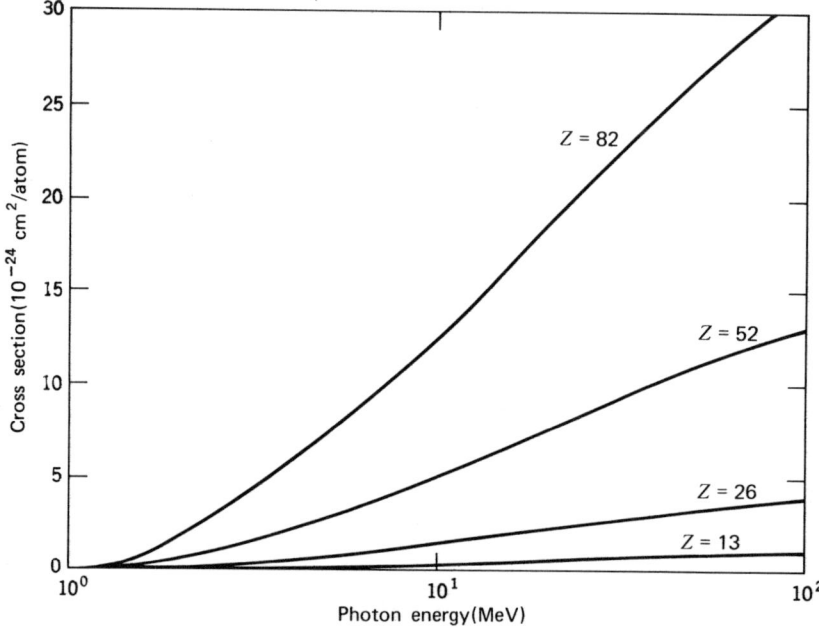

Fig. 4-4 Cross sections for the pair production process for representative materials as a function of photon energy. (From K. Z. Morgan and J. E. Turner, eds., *Principles of Radiation Protection*, John Wiley and Sons, 1967).

4-6 PENETRATION OF PHOTONS IN MATTER

In all the photon interactions described above the photon loses a large part of its energy in a single encounter. This implies an important distinction between the way in which X-rays or gamma rays penetrate matter, as compared to charged particles. As discussed earlier, charged particles make a large number of small energy losses in penetrating matter; while photons make a small number of large energy losses in penetrating matter. Therefore, if we examine the penetration of the two types (charged and uncharged) of radiation through "thin" and "thick" abosrbers we see the situation shown in Fig. 4.5. In the case of charged particles, Fig. 4-5a, thin absorbers remove some energy from all of the particles, and thick absrobers, Fig. 4-5b, will remove more energy but *all* of the particles may emerge. Of course, when the thickness becomes greater than the range of the charged particles no particles will emerge. In the case

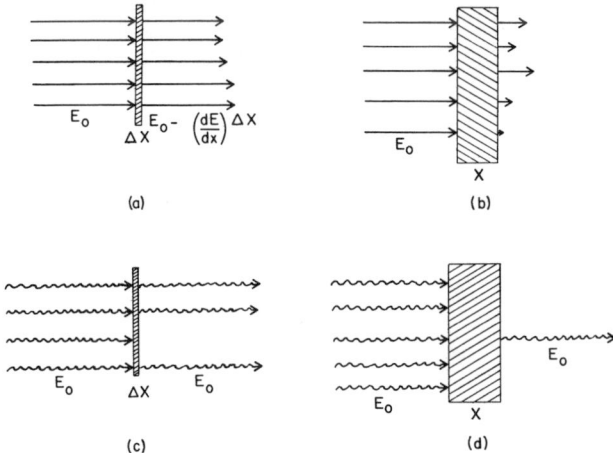

Fig. 4-5a The penetration of charged particles through a thin target. The length of the arrows is proportional to the particle energy. After penetrating a thin target of thickness ΔX all particles have nearly equivalent energies, namely $E_0 - (dE/dX)\Delta X$.

Fig. 4-5b After penetrating a target whose thickness is nearly equal to the range of charged particles all emerging particles have energies much less than E_0. Considerable spread in the particles energies is observed, and is known as "straggling."

Fig. 4-5c When X-rays or gamma rays penetrate a thin target a small fraction of the photons are absorbed. Those which emerge have the incident energy E_0.

Fig. 4-5d When X-rays or gamma rays penetrate a thick target, a large fraction of the photons are absorbed. The emerging photons have the energy E_0. Secondary electrons or scattered photons are not shown for clarity.

of photons, thin absorbers, Fig. 4-5c remove a few photons completely, but those which do not interact emerge with the initial energy $h\nu_0$. When photon interactions occur in the Compton scattering process some photons emerge at various angles with reduced energies. When the absorber is thick, fewer photons are transmitted and these have a distribution of energy resulting from one or more Compton scatterings. Indeed, even if the absorber is very thick some photons may emerge with energy $h\nu_0$, but the number transmitted would be very small. In Fig. 4-5 secondary electrons and scattered photons are not included.

4-7 ABSORPTION COEFFICIENTS

As seen from the above discussion, ranges as defined for charged particles do not apply to photon penetration into matter. Instead, we discuss photon penetration in terms of the absorption coefficient, defined as follows. In Fig. 4-6 a large number of photons, N, are incident on an absorber of thickness ΔX. Provided that ΔX is sufficiently small the number of photons interacting in ΔX is proportional to ΔX and to N. Thus we write

$$\Delta N = -\mu N \Delta X \qquad (4\text{-}6)$$

where ΔN is the change in the number of photons in the beam and the constant of proportionality, μ, is called the *linear absorption coefficient*. The preceding discussion is similar to that used in introducing cross section. Indeed, absorption coefficients and cross sections express the same physical concepts and are, therefore, mathematically similar. Equation 4-6 has the same form as the equation for radioactive decay 2-7, and we may write

$$N = N_0 e^{-\mu X} \qquad (4\text{-}7)$$

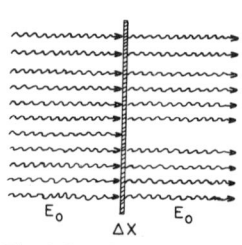

Fig. 4-6 Linear absorption coefficient. The probability that a photon will interact with the material in a volume of thickness ΔX is $\mu \Delta X$, where μ is the linear absorption coefficient.

where N_0 is the number of photons entering the absorber $(X = 0)$. This equation gives the number of photons which remain in the beam after traversing a thickness X. It is not applicable to the Compton scattered photons, but even in that case it gives the number of photons which *have not been scattered* (primary photons).

In practice, photon beams are characterized by *intensity* rather than the number of particles. The intensity of a beam is defined as the energy per unit time crossing a unit area at right angle to the beam. Then the intensity of a photon beam is the energy, $h\nu$, of each photon multiplied by the number of photons crossing the unit area in a unit time. A *flux*, ϕ, of photons is defined as the number of particles crossing in a unit time a unit area placed at right angle to the beam. Hence

$$I = h\nu\phi \tag{4-8}$$

and by analogy with Eq. 4-7

$$\phi = \phi_0 e^{-\mu X} \tag{4-9}$$

and

$$I = I_0 e^{-\mu X} \tag{4-10}$$

The latter equation is frequently employed in biological or radiation protection work since the intensity of a beam is more closely related to its ability to deposit energy in a material of interest.

Before giving practical values for absorption coefficients, that permit the calculation of beam intensity as a function of the absorption distance, X, we discuss the various *types* of absorption coefficients which have come into use. First, we make a more careful statement of the meaning of absorption coefficients, in the case where photons may be both truly absorbed in a medium or merely *scattered* from the beam. The important case in this respect is the Compton effect. Imagine a very narrow photon beam incident on a slab of absorbing material and a small detector placed a great distance away from the material, as in Fig. 4-7 *a*. In this case all photons which are scattered out of the beam are *not* recorded by the detector just as those photons truly absorbed in the material are not recorded. Such a scattering arrangement is called a "good" geometry arrangement, and absorption coefficients obtained in "good" geometry experiments or by means of "good" geometry calculations are called *total* absorption coefficients. In Fig. 4-7*b* we depict a case where the photon beam is broad and where the detector is close to the irradiated material. In this "poor" geometry case the detector measures the scattered as well as unscattered radiation, and so the measured intensity is greater than in the "good" geometry case, and the absorption coefficient is smaller. This smaller absorption coefficient corresponds more closely to the radiation actually absorbed in the material (i.e., converted to electronic kinetic energy), and is therefore called the true absorption coefficient. The total absorption coefficient less the true absorption coefficient is known as the *scatter* absorption coefficient.

It is perhaps unfortunate that other qualifying terms must be appended to the above absorption coefficients, depending on the way in which the

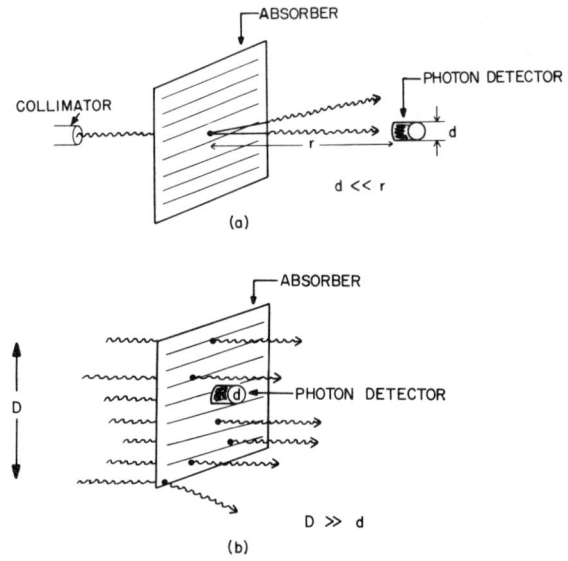

Fig. 4-7a Illustration of "good" scattering geometry. Nearly all of the scattered photons miss the detector.

Fig. 4-7b Illustration of "poor" scattering geometry. When a detector is placed near an absorber and when the beam is broad (i.e. $D > d$), it will detect scattered photons. Even if some of the photons headed for the detector should be deflected by the absorber they will strike the detector, unless they are scattered through very large angles. Furthermore, there is a small probability that photons scattered from all regions of the absorber will reach the detector.

absorption distance is specified. The absorption distance is variously specified in terms of length (cm), length times density of material (g cm^{-2}), length times the number of atoms per cm^3 (atom cm^{-2}), or length times the number of electrons per cm^3 (electrons cm^{-2}). Accordingly, the absorption coefficients are qualified by the terms: linear, mass, atomic, and electronic. However, the most commonly used ones are the linear and mass absorption coefficients, with units of cm^{-1}, and cm^{2}g^{-1}, respectively.

4-8 APPLICATION OF ABSORPTION COEFFICIENTS

Mass absorption coefficients for H_2O, Al and Pb are shown as a function of photon energy in Figs. 4-8, 4-9, and 4-10, respectively. In these figures the symbol for the total mass absorption coefficients is μ/ρ, and the symbol for the true mass absorption coefficient is μ_a/ρ. Let us illustrate the use of these curves by calculating the absorption of 1 MeV photons in aluminum. We may ask first: what fraction of the photon intensity is transmitted, which is neither absorbed in the material nor scattered from the beam, if the thickness of the aluminum is 10 cm? The fraction we seek is $I/I_0 = e^{-\mu X}$ (see Eq. 4-10). From Fig. 4-9 we see that the total mass absorption coefficient at 1 MeV for aluminum is 0.07 cm²g⁻¹. Now X must have the units of g cm⁻² in order that μX is dimensionless. Thus, $X = 10$ cm × 2.7 g.cm⁻³ = 27 g cm⁻² and $\mu X = 0.07 \times 27 = 1.89$. We find then that $I/I_0 = e^{-1.89} = 0.15$. Thus, only 15% of the photon intensity passes through the 10 cm aluminum slab without interaction. We may illustrate further by asking, what fraction of the same photon beam, passing through the same slab, is *not truly absorbed*? In this case we must use the true absorption coefficient, 0.03 cm²g⁻¹. Thus $\mu X = 0.03 \times 27 = 0.81$, and $I/I_0 = e^{-0.81} = 0.44$, or 44% of the intensity remains in the scattered plus unscattered beam.

The above approximate procedure for the calculation of photon penetration through various materials has considerable practical application. Data on absorption coefficients form the basis for calculating the energy deposition in a person exposed to X-rays or gamma radiation. It also provides information needed to calculate the filtration required in medical X-ray units. For example, we see immediately from Fig. 4-8 that a large fraction of the total intensity of 10 keV (0.01 MeV) X-rays is absorbed in the first few mm of tissue or H_2O. This means that a dental X-ray machine which emits a considerable number of 10 keV X-rays cannot be effective in producing a radiogram of the teeth, yet the energy ab-

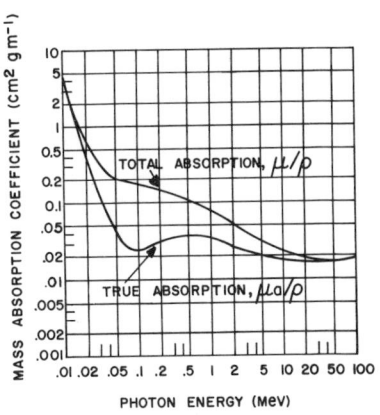

Fig. 4-8 Mass absorption coefficients in water as a function of photon energy.

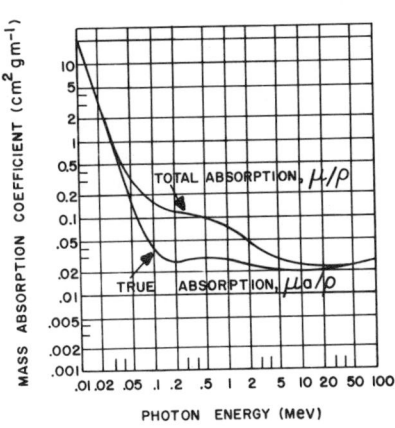

Fig. 4-9 Mass absorption coefficients for aluminum (density, $\rho = 2.70$ g cm⁻³) as a function of photon energy.

Interaction of X-Rays and Gamma Rays with Matter

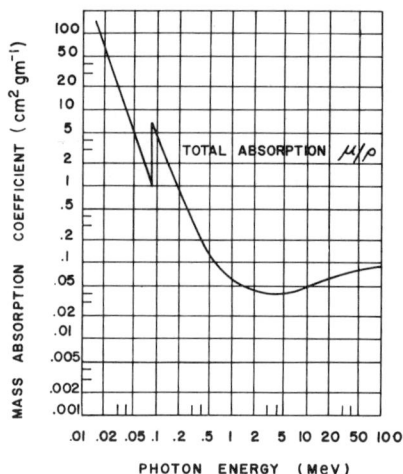

Fig. 4-10 Mass absorption coefficient for lead (density, $\rho = 11.35$ g cm^{-3}) as a function of photon energy.

sorbed in the overlaying tissues may be a radiation hazard. Fortunately, this situation is easily alleviated by placing suitable filters over the exit aperature of the X-ray machine. We see from Fig. 4-9 that a few millimeters of aluminum will remove most of the low energy (soft) X-rays, while permitting the higher energies to pass through.

Perhaps the most practical application of the data shown in Fig. 4-10 for lead is in the shielding of X-rays or gamma rays. Notice that the mass absorption coefficient for lead is larger over the entire range of energy than for water or aluminum. This means that, per unit of weight, lead is more effective in shielding against X-rays or gamma radiation. For this reason, and because lead is a common and inexpensive material, it is almost universally used where photon shields are needed. It is used to shield X-ray rooms, in aprons to shield physicians, technicians, and patients, and as material for containers used to transport radioactive sources.

PROBLEMS

1. Suppose that 10^4 photoelectrons are observed when 10^7 photons at a certain energy are incident normally to a thin slab. If the slab thickness is 0.1 cm and contains 4×10^{21} atoms/cm^3 calculate the cross section for the photoeffect.
[*Ans.* 2.5×10^{-24} cm^2]

2. The first electronically excited state in helium is 19 V and the ionization

potential is 24 V. Will a helium atom in its ground state absorb a photon whose wavelength is 1000 Å? Explain. What will be the electron energy when the helium atom is ionized by the absorption of a photon whose wavelength is 400 Å? [*Ans.* No, 7 eV]

3. A photon whose wavelength is 500 Å creates a photoelectron in a gas whose ionization potential is 15 V. Calculate the kinetic energy of the ejected electron in electron volts. Can the electron excite electronic states in the atomic gas, if these states are 5 V above the ground state?

4. A photon whose wavelength is 100 Å interacts in a photoelectric process with a valence electron whose binding energy is 6 eV. Calculate the kinetic energy of the ejected electron.

5. An inner electron is bound with 600 eV. What would be the energy of this electron, if involved in a photoeffect with 10 Å X-rays? [*Ans.* 634 eV]

6. What is the probability that a 1 MeV photon will create a photoelectron in passing normally through a slab of lead whose thickness is 10^{-2} cm? The data in Fig. 4-2 or Fig. 4-10 may be used. [*Ans.* 7.7×10^{-3}]

7. It is found that a certain material emits photoelectrons only for light of wavelength less than 5000 Å. Determine the kinetic energy of an electron ejected from this material by a 15 eV photon.

8. Use Figs. 4-2 and 4-3 to compare the photoelectric and Compton cross sections for aluminum, iron, and lead at 0.1 MeV and at 1.0 MeV.

9. Calculate the minimum photon energy required to produce a proton and an antiproton. [*Ans.* 1.9 GeV]

10. A 10 MeV photon interacts with matter in a pair-production process. Calculate the kinetic energy of each particle. A photon whose wavelength is 1 Å interacts photoelectrically with an electron in a material whose binding energy is 1 keV. Calculate the kinetic energy of the ejected electron.

11. A 2 MeV photon creates an electron-positron pair. Assuming each created particle to have equal energy, what thickness of aluminum is needed to stop the particles? [*Ans.* 0.08 cm]

12. Calculate the linear absorption coefficient corresponding to Fig. 4-5 c when ΔX is 1 cm. Assume each arrow represents a large number of photons. Calculate two different ways: thin slab approximation and the exponential absorption formula. Why are the answers different?
[*Ans.* 0.25 cm^{-1}, 0.30 cm^{-1}]

13. Derive an expression for the ratio σ/μ, where σ is the cross section in cm^2 for a process and μ is the linear absorption coefficient for the same process in cm^{-1}. Check the results by a dimensional analysis.

14. For 100-keV X-rays the true absorption and the total absorption coefficients in lead are both equal to 5 cm^{2}g^{-1}. What is the thickness (in cm) of lead re-required to reduce the transmitted intensity to 0.050 times the original value? (The density of lead is 11 gcm^{-3}). Explain why Compton scattering can be neglected when the true absorption and total absorption coefficients are nearly equal.

15. A 0.50 MeV gamma-ray beam is incident on an aluminum absorber of thickness 10 g cm^{-2}. The total absorption coefficient is 0.1 cm^{2}g^{-1} and the true absorption coefficient is 0.03 cm^{2}g^{-1}. Calculate the fraction of the photon beam intensity which is transmitted without interaction.

[*Ans.* 0.368]

16. A 10-keV X-ray has an absorption coefficient equal to 5 cm^{2}g^{-1} in water. At this energy the true absorption and total absorption coefficients are equal. What fraction of the radiation intensity is transmitted through 2 mm of tissue (water)? What fraction of the transmitted radiation is Compton scattered?

17. A uniform, parallel, beam of 1 MeV photons is incident normally to a 20 cm slab of water. What fraction of the beam intensity is truly absorbed? What fraction is scattered? What fraction is transmitted without interaction?

[*Ans.* 0.50, 0.25, 0.25]

18. A dentist places the nose cone of his 100-kVp X-ray machine against the face of a patient to obtain an X-ray picture of the teeth. Without filtration considerable low-energy (assume 20 keV) X-rays are emitted. If the intervening tissue is equivalent to 5 mm of water, calculate the fraction of the 20 keV intensity absorbed by the tissue. What thickness of aluminum would be required to reduce the 20 keV radiation exposure by a factor of 10? Calculate the intensity transmitted by the aluminum filter at 100 keV. On adding the filter, must the exposure time be increased to obtain the same quality of X-ray picture?

[*Ans.* 0.30, 3 mm, 0.86 I_0, No]

CHAPTER FIVE

Interaction of Neutrons with Matter

In an earlier paper I showed that the radiations excited in certain light elements by the bombardment of α-particles consist, at least in part, of particles which have a mass about the same as that of the proton but which have no electric charge. These particles, called neutrons, have some very interesting properties. The most obvious properties of the neutron are its ability to set in motion the atoms of matter through which it passes and its great penetrating power.

J. Chadwick

5-1 INTRODUCTION

A central theme of this book — the ability of radiation to penetrate matter and to deposit energy at great depths — cannot be presented in a simple way in the case of neutrons. The penetration of charged particles was described with the simple concept of range, while the penetration of electromagnetic radiation was described with the exponential absorption law. For electromagnetic radiation we had a hint of a difficulty since the

Compton effect required the use of two types of absorption coefficients. For neutrons the various interactions are so complex that it is advisable to postpone the discussion of penetration into matter until the end of Chapter 7, when the concept of radiation dose will facilitate the discussion.

Neutrons have no charge and carry no electric field, therefore they penetrate the electronic cloud of an atom. Neutrons interact with the nucleus by means of very strong, short-range nuclear forces. Many types of nuclear interactions have been discovered, including capture processes in which electromagnetic radiation is emitted, processes in which charged particles are released, and the splitting of the nucleus in fission processes. Such reactions strongly depend on the type of atomic nucleus, hence the isotope effect is very important. The scattering of neutrons by the nucleus is a less specific but common process, which contributes appreciably to the total cross section.

Neutron interactions with the nucleus depend a great deal on the neutron energy. *Thermal neutrons* are those neutrons which have energy characterized by the ambient temperature. Neutrons may collide with molecules (e.g., those of air) at a given temperature and eventually reach the same distribution of energies as the molecules with which they collide. At room temperature, therefore, their average energy is 0.038 eV, and the most probable energy in the distribution of energy is 0.025 eV. *Fast neutrons* are generally defined as those with energy greater than 10 keV but less than 10 MeV. *Intermediate energy neutrons* have energies in the range between thermal and fast, while *relativistic neutrons* are those with energy exceeding 10 MeV.

5-2 INTERACTION OF THERMAL NEUTRONS WITH MATTER

Thermal neutron interactions can be usefully classified as *radiative capture*, *charged particle*, and *fission interactions*. A compound nucleus, which decays into observed final products, may be formed by neutron capture.

A classic example of the radiative capture interaction involves neutron capture by hydrogen to form stable deuterium, that is,

$$^{1}_{0}n + {}^{1}_{1}H \rightarrow {}^{2}_{1}H + \gamma \tag{5-1}$$

in which a 2.2 MeV gamma ray is emitted and occurs with a cross section equal to 0.33 barns. Another famous capture gamma process involves thermal neutron capture by uranium-238, that is,

$$^{1}_{0}n + {}^{238}_{92}U \rightarrow {}^{239}_{92}U + \gamma \tag{5-2}$$

where the $^{239}_{92}$U isotope is radioactive and transforms by β^- decay to an isotope of neptunium, $^{239}_{93}$Np. Many isotopes useful in medicine, research, and industry are produced by radiative capture processes. Since the capture process leads to an isotope of the original material, the chemical properties of radioactive materials are nearly the same as those of their stable counterparts.

Neutron capture frequently leads to the production of charged particles. Protons are produced in the reaction leading to radioactive $^{14}_{6}$C, that is,

$$^{1}_{0}n + ^{14}_{7}N \rightarrow ^{14}_{6}C + ^{1}_{1}H \tag{5-3}$$

where the proton energy is 0.6 MeV; the cross section is 1.75 barns. Alpha particles are produced with enormous cross section (3800 barns) when thermal neutrons interact with $^{10}_{5}$B, that is,

$$^{1}_{0}n + ^{10}_{5}B \rightarrow ^{7}_{3}Li + ^{4}_{2}He \tag{5-4}$$

The Q-value for this process is 2.8 MeV; the respective energies of the two charge particles can be calculated with the methods of Section 2-5.

When thermal neutrons are captured by an isotope of a very heavy element, for example, $^{235}_{92}$U, the compound nucleus may split into two particles of nearly equal masses. For $^{235}_{92}$U a *typical* fission process is

$$^{1}_{0}n + ^{235}_{92}U \rightarrow ^{147}_{57}La + ^{87}_{35}Br + 2^{1}_{0}n \tag{5-5}$$

but, in fact, the process of fission may occur in about 30 different ways. Each fission process causes the transformation of approximately 0.2 amu of mass to energy. The total fission energy is thus about 200 MeV, which is distributed among kinetic energy of fission fragments, neutron energy, neutrino energy, gamma ray energy, and beta decay energy. The cross section for fission of $^{235}_{92}$U by thermal neutrons is about 580 barns, for $^{233}_{92}$U it is 525 barns, and for $^{239}_{94}$Pu it is 747 barns.

In the region of thermal energy neutron cross sections depend strongly on neutron energy; frequently the cross sections vary inversely as the neutron velocity. All of the cross section values quoted above for thermal neutrons correspond to the neutron velocity, 2200 m/sec, which is representative of thermal velocity.

The thermal neutron interactions shown in Eqs. 5-1 and 5-3 are the important interactions occuring when thermal neutrons are absorbed in living systems. Neutrons cannot ionize or excite matter directly, but the electromagnetic radiation (e.g., Eq. 5-1), or the charged particles (e.g., Eq. 5-3) they produce do ionize and excite matter as explained in Chapters 4 and 3, respectively.

5-3 INTERACTIONS OF FAST NEUTRONS WITH MATTER

Fast neutrons, with energies in excess of 10 keV, interact with matter in a variety of ways. Of greatest interest to radiation protection and medical application are the cross sections of fast neutrons with the principal elements H, N, C and O. The total cross sections for these elements are shown in Fig. 5-1. By far the major contribution to the total cross sections in this range of energy is elastic scattering. Elastic collisions are collisions in which both momentum and kinetic energy are conserved.

From the basic conservation laws it is easy to calculate the energy given to various types of elements when collisions are elastic (see Fig. 5-2). It may be shown that the maximum energy given to a "recoiling" atom is

$$E_{\max} = \frac{4Mm}{(M+m)^2} E_0 \tag{5-6}$$

where m is the neutron mass, M is the mass of the struck atom and E_0 is the neutron energy before collision. When neutron collisions are viewed from the center of mass of the two particles, usually scattering in any direction is equally probable (isotropic scattering). Under this condition the average energy of a recoil atom is given by

$$E_{av} = \frac{2Mm}{(M+m)^2} E_0 \tag{5-7}$$

The average energy given a recoil proton is $\frac{1}{2}E_0$ and the maximum energy

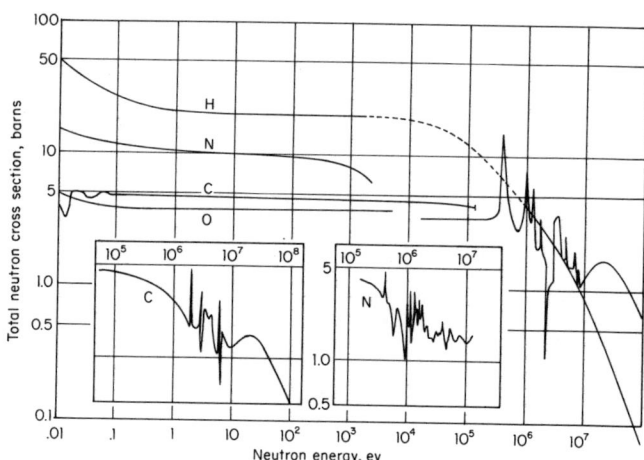

Fig. 5-1 Total neutron cross sections as a function of neutron energy for hydrogen, nitrogen, carbon, and oxygen.

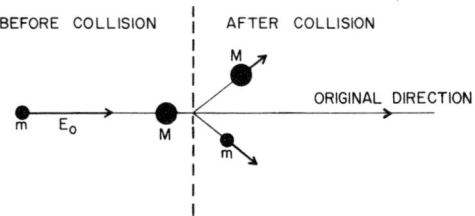

Fig. 5-2 Illustration of the elastic collision of a fast neutron (mass, m) with an atom (mass, M) at rest. Both energy and momentum are conserved in the collision.

is E_0. Both the average and maximum energy given to heavy recoils, that is, energetic C, N, and O atoms, is much less than in the proton case as easily seen from Eqs. 5-6 and 5-7.

Because neutrons lose a large fraction of their energy when colliding with the H atom, hydrogen containing materials, for example, H_2O, are used to shield against fast neutrons. Since hydrogen itself captures neutrons when moderated to thermal energy, see Eq. 5-1, the energetic gamma ray can produce health hazards. For this reason neutron shields are often made of water to which a boron component (e.g., borax,) is added so that less penetrating particles (see Eq. 5-4) are produced instead of the gamma ray.

5-4 INTERACTIONS USEFUL IN NEUTRON DETECTION

An interaction with a cross section known over a range of energy and which leads to a detectable reaction product can be used to detect neutrons. However, interactions in which the cross sections behave in specific ways are most useful. For example, the total number of neutrons, independent of their energy, can be determined when the cross section is constant, or independent of energy. A cross section which has a threshold behavior, that is, it rises rapidly above a critical energy (threshold energy) and remains nearly constant above that energy is useful in determining the number of neutrons above the threshold energy.

Cross sections of the type described above are shown in Fig. 5-3. The cross sections for $^{239}_{94}Pu$, $^{237}_{93}Np$, and $^{238}_{92}U$ are fission cross sections, hence neutron detection involves measurement of the fission product radioactivity left in a foil after previous exposure to fast neutrons. The $^{32}_{16}S$ reaction is the charged particle type, that is,

$$^{1}_{0}n + ^{32}_{16}S \rightarrow ^{32}_{15}P + ^{1}_{1}H \tag{5-8}$$

where ^{32}P is radioactive and emits detectable β^- radiation.

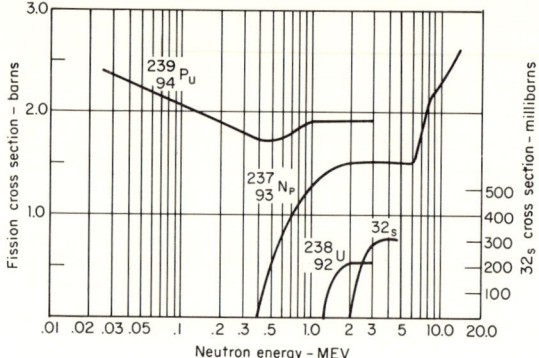

Fig. 5-3 Cross section as a function of neutron energy (for listed elements).

Note should be made of the fact that the fission cross section of $^{239}_{94}$Pu rises quite strongly to 747 barns in the thermal region, thus relatively few thermal neutrons would obscure the detection of fast neutrons. However, the $^{239}_{94}$Pu foil can be exposed to neutrons when it is shielded with $^{10}_{5}$B to filter thermal neutrons. When this is done the effective fission cross section is

$$\sigma_{eff} = \frac{N}{N_0} \sigma_F \qquad (5\text{-}9)$$

where N/N_0 is the fraction of neutrons reaching the foil and σ_F is the fission cross section, and both factors depend on neutron energy. The factor N/N_0 can be calculated from the known cross section of $^{10}_{5}$B. Thus

$$\frac{N}{N_0} = \exp\left(-\sigma_B l N_B\right) \qquad (5\text{-}10)$$

where σ_B is the boron cross section, l is the thickness of the boron shield and N_B is the number of boron atoms ($^{10}_{5}$B) per cm³. Figure 5-4 shows the calculated values of σ_{eff} for various values of the boron shield thickness (l) and an assumed $^{10}_{5}$B density of 1.11 g/cm³. As seen in Fig. 5-4, the effective ^{239}Pu fission cross section represents a very desirable type of detector with variable thresholds.

The above are examples of the use of solid foils which are useful for determining the number of neutrons in the fast energy region. Thermal neutrons are frequently measured by using gold foils in the capture reaction forming $^{198}_{79}$Au which is radioactive with a 2.69 day half-life, decaying to $^{198}_{80}$Hg by β^-(0.97 MeV) and γ(0.411 MeV) emission. The boron capture leading to charged particles (Eq. 5-4) is also commonly used to measure thermal neutrons. In this case gaseous BF_3 is used directly in gas filled detectors to count the number of neutrons captured.

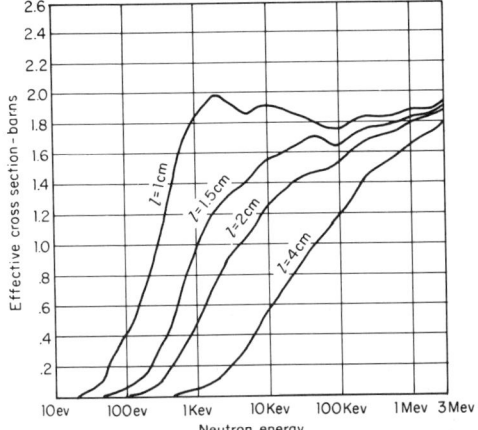

Fig. 5-4 Effective fission cross section of ^{239}Pu when shielded with indicated amounts of $^{10}_{5}$B.

PROBLEMS

1. A neutron has a velocity equal to 2×10^6 m/sec. Classify it according to the energy subdivisions given at the end of Section 5-1.

2. Calculate the gamma ray energy emitted in the process described in Eq. 5-1. Table 2-1 contains the necessary atomic masses. [*Ans.* 2.2 MeV]

3. Show that, when a thermal neutron interacts according to Eq. 5-3, the kinetic energy of the recoiling ^{14}C atom is $(M_H/M_C) E_H$ where E_H is the kinetic energy of the recoil proton.

4. Calculate the energies of the charged particles $^{7}_{3}$Li and $^{4}_{2}$He in the reaction of Eq. 5-4, assuming $Q = 2.8$ MeV. [*Ans.* 1.0 MeV, 1.8 MeV]

5. If the energy released per fission in Eq. 5-5 is 200 MeV, calculate the loss of mass.

6. When particles of the same mass are elastically scattered in Fig. 5-2, show that the sum of the scattering angles is 90°. This is most easily seen by drawing a triangle representing momentum conservation.

7. Show that the energy of a neutron scattered at an angle θ is the initial energy times $cos^2 \theta$, when the scattering process is an elastic collision with hydrogen atoms.

8. When a beam of fast neutrons is incident normally on a water target, a proton scatters straight ahead with 4 MeV kinetic energy. What is the neutron energy? What energy would the oxygen atom acquire if it is scattered straight ahead? [*Ans.* 4 MeV, 0.89 MeV]

9. Calculate the average energy given to a carbon atom when it is elastically scattered with 2 MeV neutrons.

10. Fast neutrons collide with water and make elastic collisions with both H and O. Calculate the ratio of the average energy imparted to oxygen (atomic mass 16) compared to the average energy imparted to hydrogen.

11. A 1 MeV beam of neutrons is incident normally on a plane slab of water whose thickness is 1 in. Calculate the fraction of neutrons which scatter with hydrogen, neglecting scattering with oxygen. The neutron-hydrogen scattering cross section is 4 barns at 1 MeV. [*Ans.* 0.49]

12. If the cross section of the neutron capture process in $^{10}_{5}B$ is 3800 barns at a neutron velocity of 2200 m/sec, calculate the cross section at 1 eV, assuming that the cross section obeys a 1/v law. [*Ans.* 600 barns]

13. Calculate the probability *a* neutron whose energy is 100 eV will penetrate a 1 cm shield of $^{10}_{5}B$ whose density is 1.11 g/cm³. Use the method of Problem 12 for the neutron cross section. Calculate the effective cross section for $^{239}_{94}Pu$ fission at 100 eV when a Pu foil is placed inside this shield. Assume the $^{239}_{94}Pu$ fission cross section is 23 barns at 100 eV. [*Ans.* 0.018, see Fig. 5-4]

14. A neutron beam is normally incident on a "thin" slab of carbon atoms. Let the scattering cross section be σ and the density of carbon atoms be ρ_c (atoms cm⁻³). What is meant by a "thin" slab? What is the probability that a neutron will be scattered?

CHAPTER SIX

Absorbed Energy and Its Measurement

I often say that when you can measure what you are speaking about, and express it in numbers, you know something about it.

Lord Kelvin

6-1 INTRODUCTION

In this chapter we discuss absorbed energy and its measurement. Absorbed energy is the physical quantity most often correlated with biological effects of radiation. The stopping power of a medium for charged particles is closely related to the amount of energy absorbed in the medium. However, the relationship often is complicated by the fact that secondary particles may carry some of the energy lost by the primary charged particle out of the small region where the interaction occurred. Furthermore, charged particles moving with relativistic velocities may lose energy by the production of electromagnetic radiation (i.e. *Bremsstrahlung*). Particle energy, fluence, and stopping power, are all

6-2 SOURCES AND SOURCE STRENGTH

Source strength is the number of particles of a specified type emitted from a source per unit of time; an alpha source which emits 10^6 alpha particles per second is a source whose strength is 10^6 sec^{-1}. The strength of a radioactive source is closely related to its *disintegration rate*. We show below that these two quantities are not identical.

In Fig. 6-1 is shown the *decay scheme* for ^{60}Co, which has a half-life of 5.26 years. As the figure shows, ^{60}Co may decay to stable ^{60}Ni by two routes. The most probable route (almost 100% of the cases) is one in which a beta particle (β_1^-) is given off with a maximum energy of 0.314 MeV, followed by a gamma ray (γ_1) with 1.173 MeV, and a second gamma ray (γ_2) of 1.332 MeV. Only rarely (0.01% of the cases) does emission occur by the second route in which a beta particle (β_2^-) is given off with a maximum energy of 1.48 MeV and a gamma ray (γ_2) with an energy of 1.332 MeV. Neglecting the second route, we may easily compare source strength and disintegration rate. Each time a disintegration occurs one beta particle (β_1^-) and two gamma rays (γ_1, γ_2) are emitted. If we define source strength as the *total* number of particles emitted per second, then the source strength is three times the disintegration rate. On the other hand, the beta particle source strength is just equal to the disintegration rate. The gamma ray source strength is approximately two times the disintegration rate.

The disintegration rate is involved in a special definition. The *curie* is defined as that *amount* of a radioactive material which has an activity of 3.70×10^{10} distingreations per second. Madame and Pierre Curie's work with radium is thus fittingly recognized, inasmuch as 1 g of radium experiences approximately 3.7×10^{10} disintegrations per second.

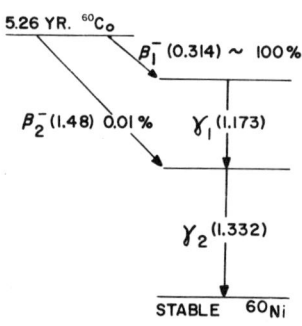

Fig. 6-1 Simplified decay scheme of ^{60}Co. Numbers in parentheses are energies in MeV.

6-3 THE CONCEPT OF SOURCE SPECTRA

In Section 6-2 we learned that each time the ^{60}Co nucleus decays to stable ^{60}Ni one beta-particle and two gamma-rays are given off. The energy of the beta-particle listed in decay schemes is the energy of the most energetic beta-particle; other beta-particles are emitted which are less energetic. In fact it is found that some beta-particles from ^{60}Co have nearly zero energy and some have the maximum energy of 0.314 MeV, and beta-particles of all intermediate energies are observed. When we plot the *number* of particles observed per unit of energy-interval as a function of *energy* a continuous function (called the particle spectrum of the beta source) is found. The particle spectrum of ^{60}Co beta particles is illustrated in Fig. 6-2. The product of the ordinate value for a given abscissa, at energy E, times a small range of energy ΔE gives the relative number of particles emitted in the energy interval ΔE.

The spectrum of all types of radiation can be constructed in a similar way, however, the function may not be smooth and continuous; in this case the term "bar diagram" or "line spectrum" is usually used instead of the term "spectrum." For example, if the number of gamma rays emitted from ^{60}Co for an arbitrary number of disintegrations is plotted as a function of energy, two vertical lines, at the energies 1.173 MeV and 1.332 MeV, are found. Figure 1-9 is an example of both a continuous and line spectrum obtained from an X-ray tube.

Charged particles are deflected by electric or magnetic fields, and a measure of the magnitude of these fields required to produce a fixed deflection depends on the energy and mass of the particle. Therefore, the

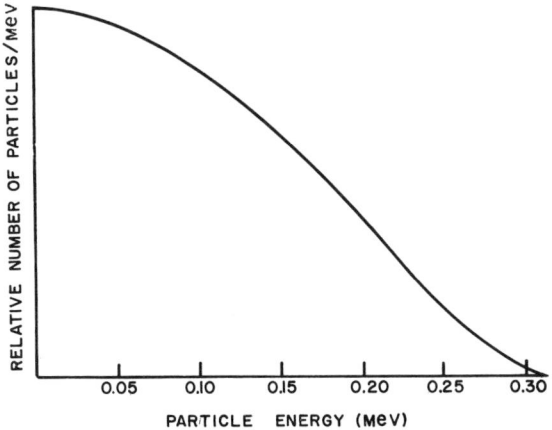

Fig. 6-2 Particle spectrum for ^{60}Co beta.

spectrum of particles of known mass, such as beta-particles, protons, or alpha-particles can be determined by observing the number of particles deflected into a detector as a function of the field strength. The energies of uncharged particles such as neutrons, and electromagnetic radiation, cannot be examined in this way. However, their charged particle secondaries, whose energies are simply related to the energies of the uncharged particles, may be analyzed by deflection methods.

Many other techniques have been developed for spectral analysis of radiation sources. Measurement of total energies of particles absorbed in detectors is the basis of other spectroscopic methods. Frequently these detectors are ionization chambers, proportional counters (Sections 6-7 and 6-8), light-producing crystals such as NaI, and solid state charged particle detectors (Section 6-6). Nuclear reactions which occur at well defined "threshold" energies are particularly useful in investigating the energy spectrum of neutrons (Section 5-4).

6-4 FLUX AND FLUENCE DUE TO POINT SOURCES

As pointed out in the introduction to this chapter, the flux of particles at a point determine the time rate of energy absorbed in a region around the point for either charged or uncharged particles. The flux, in turn, is governed by the source strength, the distance of the point from the source and the type and amount of material between or around the source or the point.

Imagine a point-source, S, and a point P (as in Fig. 6-3), which are separated by a distance R. If no interactions occur between the radiation and matter (e.g., in a vacuum) the flux, ϕ, is given by the *inverse square law*, that is,

$$\phi = S/4\pi R^2 \qquad (6\text{-}1)$$

where S is the source strength. This simple equation follows immediately from the conservation of particles. If S has units of particles per second

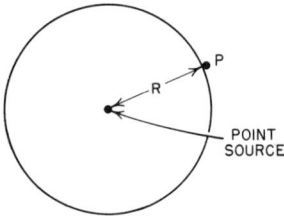

Fig. 6-3 Inverse square law. The particle flux is $\phi = S/4\pi R^2$, where S is the source strength.

(denoted by sec^{-1}) and if R has units of cm, then ϕ has units of particles per cm^2 per sec (denoted by cm^{-2} sec^{-1}). Particle fluence, defined as the number of particles crossing a unit of area, is equal to the product ϕt where t is the time. Thus, assuming no intervening material, we have for the fluence

$$F = \phi t = St/4\pi R^2 \qquad (6\text{-}2)$$

for a point source of constant strength.

Particle flux and fluence also depend on the type of materials interposed along the path and to a lesser degree on materials around the path because of radiation scattering. As emphasized in Chapter 4, an important distinction must be made between the way in which charged and uncharged particles penetrate through matter. Thus, for charged particles the interposition of materials along the path does not change ϕ or F appreciably if the charged particles have a *range* in the interposed material greater than the material thickness. Of course, the interposed material does reduce the energy of the transmitted particles. When the thickness is greater than the range, ϕ and F fall to zero. When materials are interposed along the path of uncharged particles, like neutrons or photons, the quantities ϕ and F decrease continuously as the thickness of the material is increased. We noted in Chapter 4 that in the case of photons, quite frequently the energy of the photons which are transmitted tends to remain unchanged. However, for fast neutrons the situation is more complex because nearly all neutrons which interact have suffered energy losses, but remain to be considered as scattered neutrons.

6-5 INTENSITY AND ABSORBED ENERGY

The *intensity of a beam of radiation* is defined as the *energy* crossing a unit of area normal to the beam in a unit time. Thus, the intensity, I, of a beam of monoenergetic particles of energy ϵ is

$$I = \epsilon\phi \qquad (6\text{-}3)$$

In Chapter 4 the intensity of a photon beam was defined in a similar way except there $h\nu$ was used instead of ϵ. When the beam is not monoenergetic we may write for the intensity

$$I = \epsilon_1\phi_1 + \epsilon_2\phi_2 + \cdots + \epsilon_n\phi_n = \bar{\epsilon}\phi \qquad (6\text{-}4)$$

where ϕ_1 is the flux of particles at energy ϵ_1, ϕ_2 is the flux of particles at

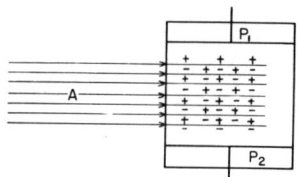

Fig. 6-4 The intensity of a beam of charged particles can be determined by measuring the amount of ionization produced in gas.

energy ϵ_2, etc. On the right-hand side of Eq. 6-4, ϕ is the total flux and $\bar{\epsilon}$ is the average energy which depends on the particle spectrum.

In Fig. 6-4, we consider a charged particle beam with a cross sectional area of A, which enters into a gas where it is totally absorbed. It has been found by experiment that for most gases the ratio,

$$W = \epsilon/n_i \qquad (6\text{-}5)$$

(where n_i is the number of *ion pairs* produced by one particle of energy ϵ) is nearly constant regardless of the value of ϵ and the type of particle. This "W-value" of a gas typically has the value of 32 eV per ion pair (i.e., 32 eV/ip). W-values are well known for a number of gases and this quantity provides a precise and convenient method for quantitative radiation measurements. For example, in our schematic problem of Fig. 6-4 we may write for the total time rate of ion pair production, $\dot{N}_i$,

$$\dot{N}_i = n_i \phi A = \epsilon \phi \frac{A}{W} \qquad (6\text{-}6)$$

and from Eq. 6-3

$$I = W\dot{N}_i/A \qquad (6\text{-}7)$$

Therefore, from a measurement of $\dot{N}_i$, combined with a knowledge of W and A, the intensity can be determined. If one applies a potential difference between the plates P_1 and P_2, as done in a *parallel plate ionization chamber*, $\dot{N}_i$ is proportional to the current which flows provided that the potential difference is sufficient to *collect* all ion pairs. More details on the ionization measurements will be given in Section 6-7.

It is easily shown that $\dot{N}_i$ is also proportional to the time rate of energy absorption by recognizing that the rate of energy absorption in the gas, $\dot{E}_{abs}$, is

$$\dot{E}_{abs} = \epsilon \phi A \qquad (6\text{-}8)$$

thus

$$\dot{E}_{abs} = W\dot{N}_i \qquad (6\text{-}9)$$

or

$$E_{abs} = WN_i \qquad (6\text{-}10)$$

In most practical cases it is difficult to insure that the idealized conditions of a parallel, uniform beam actually hold true, and it is often difficult to relate $\dot{N}_i$ to I. However, even in these cases Eqs. 6-9 and 6-10 may be used to obtain $\dot{E}_{abs}$ and E_{abs} from measurements of $\dot{N}_i$ and N_i. Fortunately, it is the energy absorbed in a biological system that produces damage in it; the beam intensity itself is of secondary importance.

6-6 MEASUREMENT OF ABSORBED ENERGY

Man is not endowed with a sense for ionizing radiation, and so he makes analysis of irradiated physical, chemical, and biological systems for practical methods of radiation detection. Sensitivity, accuracy, and versatility factors favor the physical and chemical systems, and hence only these are discussed.

Of the physical systems, one naturally thinks first of *calorimetry*, for most of the energy in a beam of radiation is eventually degraded to heat. However, it is known that if in a short time (e.g., 1 sec) radiation energy is absorbed in living tissue to the extent of 10^6 ergs/g that the tissue will be damaged beyond repair. On the other hand, this amount of absorbed energy per gram (from ionizing radiation or as heat) will cause only a 0.024° C change in body temperature! This most remarkable fact shows at once the extreme ineffectiveness of heat (compared to ionizing radiation) on living systems, and the difficulty of making use of calorimeters for routine radiation measurements. These remarks are amplified in Problem 6-7.

The independence of the W-value of a gas on particle energy implies that the amount of ionization is proportional to absorbed energy. Also, *ionization methods* are the most versatile and sensitive of the physical methods. Consider the deposition of one erg of energy in a gas by radiation which eventually leads to the production of ion pairs. Since 1 erg = 6.25×10^{11} eV the number of ion pairs produced in a gas is given by

$$N_i = \frac{6.25 \times 10^{11} \text{ eV}}{32 \text{ eV/ip}} = 1.96 \times 10^{10} \text{ ip}. \tag{6-11}$$

Thus 1 erg of energy absorbed in a gram of living tissue, a radiation exposure not producing observable biological damage, would produce in a gas about 2×10^{10} ip. Ionization detectors such as *ionization chambers*, *proportional counters*, and *Geiger-Mueller counters*, can be used to detect a few ion pairs and this accounts, in part, for the wide acceptance of ionization devices in biological and medical applications. More information on these instruments will be given in the next section.

Solid state charged particle detectors involve an interesting and very useful extension of the ionization method. In the solid, for example in single crystals grown from highly purified silicon or germanium, an energetic charged particle produces electron-hole pairs analogous to the electron-ion pairs produced in a gas. Free charge thus liberated is collected by the application of an electric field to the thin strip of solid material. Since the amount of charge which is liberated in the solid state detector is approximately proportional to the energy absorbed from the charged particle, a measurement of the collected charge can be used to determine the energy absorbed from the particle.

Certain materials, called *scintillators*, are known to emit light when radiation energy is absorbed. Much pioneering work on radioactivity was done by viewing small bursts of light emitted from zinc sulfide powder when single alpha particles were absorbed. In modern technique, the human eye is replaced with photomultiplier tubes which are used to record the number of light flashes (or events). With this quantitative technique the amount of light associated with each event can also be recorded, and thus information on the energy absorbed in the detector is obtained. Scintillation detectors are extensively used in medical "scanning" applications described in Chapter 10.

Photographic film or *plates* are blackened by the absorption of a few ergs of radiation energy per cm^2. This technique is one of the oldest methods, having been the means by which natural radioactivity was discovered by Becquerel (see Section 2-1). Small crystals of a silver halide (e.g., silver bromide) are suspended in a gelatin which is spread over a cellulose acetate film or glass plate. On exposure to radiation, some of the secondary electrons released in the medium interact with the silver halide and cause the reduction of silver atoms. The development process amounts to a chemical separation in which metallic silver is left on the film or glass plate. It is this silver deposit which accounts for film blackening; the degree of blackening is related to the absorbed energy per unit mass. The photographic film technique has been the most common method of monitoring large numbers of individuals for radiation exposure.

The photographic film technique is but an example of the *chemical method* generally. Frequently solutions are prepared, exposed to radiation, and the yield of a selected chemical is measured. For example, the number of ferrous ions which are converted to ferric ions is proportional to the energy absorbed in a solution of ferrous ammonium sulfate and sulfuric acid.

6-7 IONIZATION CURRENT METHODS FOR THE MEASUREMENT OF ABSORBED ENERGY

We have already mentioned that the remarkably constant value of W provides an accurate and sensitive method for the measurement of absorbed energy. In this section we wish to discuss in more detail the ionization methods and particularly methods of measuring the number of ion pairs produced in a gas.

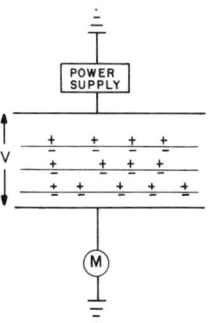

Consider first the simple parallel plate ionization chamber, Fig. 6-5. When radiation interacts with a gas it produces ion pairs of positive and negative charge. Usually the positive ion has one unit of charge and the negative ion is a free electron with a small amount of energy, typically 1 or 2 eV. Many of the electrons would *recombine* with the positive charge to form a neutral particle if left in a field-free region. However, an electric field causes a drift of the electrons in the direction of the positive plate, and a much slower drift of the heavy positive ions in the direction of the negative plate. This drift of electrical charge constitutes an electrical current which flows through the external circuit and is measured by the meter, M. If the current i is measured for various values of the applied potential difference, V (which determines the electrical field E), the relationship $i(V)$ plotted in Fig. 6-6 is obtained. At first i increases with V in a manner determined by *ion pair recombination*. When V is sufficiently large i reaches a constant value, or plateau, and it is this current i_0, which gives the *rate* of ion pair formation. When the average rate of ion pair formation is 1 ip/sec the current which flows is 1.6×10^{-19} amp. Moderately good *electrometers* can measure currents less than 10^{-15} amp, hence they can detect the formation of 10^4 ip/sec corresponding to a rate of energy absorption of about 3×10^5 eV/sec or about 4.8×10^{-7} erg/sec.

Fig. 6-5 Diagram of the ionization chamber method for measuring the rate of ion pair production.

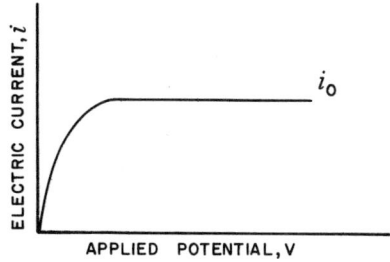

Fig. 6-6 Current, i, flowing through the external circuit of an ionization chamber as a function of applied potential, V.

6-8 IONIZATION PULSE METHODS FOR THE MEASUREMENT OF ABSORBED ENERGY

Ionization chambers operated in the plateau region may also be used to *count the number* of particles interacting with the gas. As ions move toward the collecting plates, a small drop occurs in the plate's potential which, after ion collection, is restored to the initial potential by the external battery. This time variation of the potential, (or electrical impulse, called pulse for brevity) can be amplified and recorded electronically. In this way the passage of each particle which dissipates sufficient energy can be detected. In fact the spectrum of absorbed energies can be determined in this fashion, since the size or height of each pulse is in direct proportion to the number of ions formed by each of the particles. Pulse ionization chambers have limited practical value because of electronic noise problems. Even in elaborate pulse amplifiers electronic-noise signals are about as large as the pulse height produced when a particle loses 1 MeV of energy. Pulses produced by particles having less than 1 MeV of energy cannot be observed easily.

With *proportional counters* the electronic noise problem can be greatly reduced. Proportional counters are made by stringing a fine wire down the axis of a conducting cylinder, as shown in the cross section sketch of Fig.

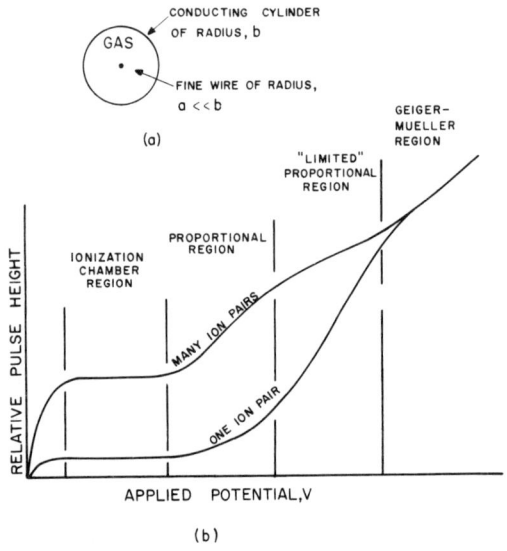

Fig. 6-7(a) Cross section of a proportional counter. (b) Relative pulse height as a function of applied potential for a cylindrical counter.

6-7. By filling with a counting gas at suitable pressure and application of sufficient voltage, the pulses in the proportional counter are many times (e.g., 10^3 or 10^4) greater than those observed in the pulse ionization chamber. Electrons are separated from the positive ions in the manner discussed above by the field E. However, the field very near the fine wire of a proportional counter is so strong that the initially low-energy electrons gain enough energy to produce additional ion pairs in the gas. This *gas amplification* process accounts for the much larger pulses produced in a proportional counter. Unless the applied potential is made too large, the final height is proportional to the initial number of ion pairs, a fact easily recalled from the name of the device. Proportional counters are practical devices for counting energy-loss events as small as a few keV. They may be used to obtain the spectrum of absorbed energies, from which the total energy absorbed or the time rate of total energy absorption can be determined.

If the potential applied to the cylindrical geometry of Fig. 6-7 is increased further, it is found that the pulse heights are increased but the same pulse height is observed regardless of the number of initial ions. Thus, *Geiger-Mueller* counters produce large pulses of the same size independently of whether the intial event produced one ion pair or a very large number of ion pairs. *Geiger-Mueller* counters are widely employed as detectors of radiation, but have limited use in measurements of absorbed energy.

PROBLEMS

1. What is the source strength of the 1.173 MeV gamma ray from 1 curie of ^{60}Co? What is the strength of gamma rays of all energies from 1 curie of ^{60}Co? What is the strength, per curie of ^{60}Co, of the 1.48 MeV beta ray?
[*Ans.* 3.7×10^{10} sec^{-1}, 7.4×10^{10} sec^{-1}, 3.7×10^{6} sec^{-1}]

2. Calculate the flux of 1.173 MeV gamma rays at 10 cm and at 100 cm from a one curie point source of ^{60}Co in a vacuum. Calculate the number of gamma rays per sec crossing spheres of 10 cm and 100 cm radius, assuming this point source to be at the centers of the spheres. Compare the two answers with the appropriate source strength in Problem 1.

3. A detector of cross sectional area 10 cm² is placed 10 m from a point source. If the detector counts 10^4 particles per second what is the strength of the source for particles of the type detected? [*Ans.* 1.25×10^{10} sec^{-1}]

4. A source of strength S_1 is turned on for a time t_1, and another source of strength S_2 is turned on for a time t_2. Derive a formula for the particle fluence at a point whose distance from source S_1 is R_1 and whose distance from S_2 is R_2.

5. When a uniform charged-particle beam of area 10 cm² enters a gas and is totally absorbed, the observed current (due to ionization of the gas) is 10^{-6} amp. Assuming the W value of the gas is 30 eV/ip, calculate the beam intensity in units of eV sec^{-1} cm^{-2}.

6. A 5 MeV alpha-particle beam, having a cross sectional area of 2 cm², enters an ionization chamber and is stopped completely. The current which flows under voltage saturation conditions is 10 μamp. If the energy loss per ion pair is 30 eV, what is the intensity of the beam of alpha particles? What is the flux? [Ans. 9.4×10^{14} eV sec^{-1} cm^{-2}, 1.88×10^{8} cm^{-2} sec^{-1}]

7. Calculate the distance that a mass of water must fall in order that it will absorb on impact 10^6 ergs of energy per gram. Compare the hazard of falling from a building with the hazard of ionizing radiation on an energy per gram basis. (Assume that a 16 ft. fall is a lethal distance.) How much energy is imparted per gram when a person takes a bath in which the body temperature rises 1°C? [Ans. 33.5 ft., 4.18×10^7 erg/g]

8. Assume that 10^6 alpha particles of 5 MeV energy are emitted each second from a thin source placed in an ionization chamber. If the particles are completely stopped in the gas whose W-value is 36 eV/ip, calculate the number of ion pairs per second, the current (amp) which flows under voltage saturation conditions, and the charge (coul) which would be collected in 1 hr.

9. A 1 MeV proton has a W-value of 30 eV/ip in a proportional-counter filling gas. Assuming that a 1 MeV recoil carbon atom has a W-value of 40 eV/ip in the same gas, calculate the ratio of the pulse heights produced by the two particles, if both particles are completely absorbed. [Ans. 4/3]

10. Explain why a Geiger-Mueller counter is not a useful device for determining the energy absorbed from charged particles by the filling gas.

11. Give a simple argument, based on electrostatics, to explain why an electron may recombine with a positive ion in a gas.

12. Assume that the W-values for protons and carbon-recoil atoms are both equal to 30 eV/ip in C_2H_4 gas. What is the maximum number of ion pairs that can be produced by a 3 MeV neutron interacting elastically with H and with C? [Ans. 10^5, 2.8×10^4]

CHAPTER SEVEN

Radiation Dosimetry

> It is extremely important that the comparison of biological effects produced by neutrons and by the better known ionizing radiations (X-rays and gamma rays) be made under proper conditions.
>
> G. Failla

7-1 INTRODUCTION

Radiation dosimetry deals with the effects of ionizing radiation on physical, chemical, or biological systems and seeks to establish useful correlations among these effects. To understand the effects of radiation on chemical or biological systems it is necessary to study the action of radiation on simple physical systems in the gas, liquid, or solid state. The mechanisms so revealed illuminate the effects of radiation in the far more complex systems. In radiation protection work, the response of instruments is used to regulate exposure of persons to radiation. This procedure requires the correlation of physical quantities (i.e., instrument response) with possible biological damage.

In the previous chapter, absorbed energy and various methods for its measurement were discussed. It was pointed out that absorbed energy, rather than beam intensity, must be closely related to the effects of radiation on physical, chemical, or biological systems. We emphasize again that it is the fraction of the radiation energy which is absorbed in a living system that can produce effects in it.

The physical quantity most frequently correlated with biological effects is the absorbed energy per unit of mass. This quantity is called the *absorbed dose*. It is not reasonable to expect that so simple a quantity would correlate with a complex sequence of events, starting with the excitation or ionization of atoms and molecules, and ending with gross clinical observations. Indeed, many qualifications must be introduced. Besides absorbed dose, the time rate at which energy is absorbed (dose rate) is important. In addition biological effect depends on the stopping power of the particles expending energy in the mass. Under otherwise identical conditions, particles with different stopping powers may produce different end results. In Chapter 8, these and other qualifications are discussed in connection with biological data.

7-2 DOSE QUANTITIES AND UNITS

The first definition of dose quantity to receive widespread use is now referred to as an *exposure* and applies only to electromagnetic radiation (i.e., X-rays or γ-rays). The concept of exposure, which led to the adoption of the roentgen (R) as its unit, was based on the ability of electromagnetic radiation to ionize air. The exposure is defined as 1 roentgen when the X-ray or gamma-ray field produces 1 electrostatic unit, esu, of positive charge and 1 esu of negative charge in 0.00129 g of air. When conversion is made from ion pairs to energy, it is found that 1 roentgen results in the absorption of 87 ergs per g of air. Calculations show that this same radiation exposure would produce about 95 ergs/g of tissue. Early workers adopted a unit of absorbed dose closely related to this number, and was called the "roentgen-equivalent physical" or rep dose.

The second dose quantity in widespread use is the absorbed dose defined above. It applies to any kind of matter and to all radiations. The unit of absorbed dose, called the *rad*, is 100 ergs/g.

As data in Chapter 8 show, the distribution of dose with stopping power, frequently called the LET distribution, is important. The abbreviation LET stands for linear energy transfer or the amount of energy transferred to the matter per unit of length along the particle path. The LET would be equivalent to dE/dX if all of the particle energy were *locally absorbed* in the material. Thus, for a proton, LET and dE/dX are nearly equal, but for a very fast electron in which part of the energy lost by the particle radiates away from the particle path, dE/dX is larger than LET.

7-3 DOSE MEASUREMENTS BASED ON IONIZATION (BRAGG-GRAY PRINCIPLE)

In Chapter 6 we introduced the idea of absorbed energy measurements in which the constant W, the energy required to produce a pair of ions in a gas, served as a connection between the amount of ionization and the amount of energy absorbed. This indeed is the concept which must be developed further in order to achieve suitable methods of radiation dosimetry.

Suppose we first consider the case where a detector (i.e., a gas and its enclosure) consists of materials which have the same atomic composition as the material in which we wish to know the absorbed dose. If the range of the secondary charged particles (e.g., electrons in the case of gamma-ray exposures, or protons in the case of fast-neutron exposures) is greater than the dimensions of the gas enclosure then it follows that some of the energy which the primary particle releases in the gas, escapes into or even through the enclosure walls. This escape of energy, illustrated in Fig. 7-1, must be considered carefully. When the walls have the same atomic composition as the gas, the primary radiation gives rise to secondary charged particles in the walls, which are the same in kind as energy as those released in the gas. Clearly, some compensation occurs; that is, charged particles arising in the wall dissipate energy in the gas and tend to compensate for the energy released in the gas but lost to the walls. W. H. Bragg, and later L. H. Gray argued that if the thickness of the wall is (1) greater than the range of the charged secondary particles but (2) not so great that attenuation of the primary beam occurs, then the compensation is almost exact.

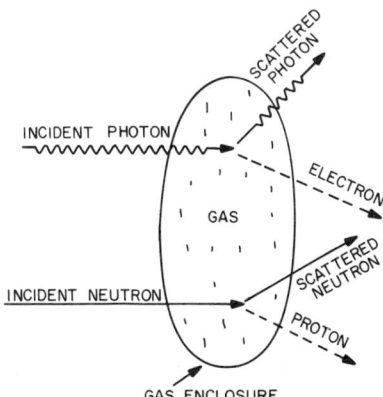

Fig. 7-1 Illustration of escape of energy released by primary particles in a volume of gas from that volume.

The *Bragg-Gray principle* states that, if a gas is enclosed with a wall of the same atomic composition, and meets the thickness conditions mentioned above, the energy *absorbed* per gram of gas can be determined by collecting and measuring the number of ion pairs produced, multiplying by the constant W and dividing by the mass of the enclosed gas. Furthermore, the energy absorbed per gram of wall, D_W, is equal to the energy absorbed per gram of gas, D_g, thus

$$D_W = D_g = N_i W / m_g \qquad (7\text{-}1)$$

where N_i is the number of ion pairs created in the gas of mass m_g. The natural units in Eq. 7-1 are eV/g. However, according to tradition D_W and D_g have units of 100 ergs/g, therefore

$$D_W(\text{rad}) = 1.6 \times 10^{-14} N_i W / m_g \qquad (7\text{-}2)$$

where W has units of eV per ion pair and m_g is in grams.

We next consider the case where the wall material has a different composition from the gas, but the same as the material for which we wish to measure the absorbed dose. The absorbed dose in the walls can then be determined, provided only that the gas cavity and its pressure are kept small, so that the secondary charged particles lose only a small fraction of their energies in the gas. However, Eq. 7-2 no longer holds true, since we must take into account the mass stopping power of the wall material relative to that of the gas for the secondary charged particles. Thus

$$D_W(\text{rad}) = 1.6 \times 10^{-14} N_i W S / m_g \qquad (7\text{-}3)$$

where S is the stopping-power ratio defined in the preceeding statement.

Finally, let us assume that the material in the radiation detector does not have the same atomic composition as the material for which we wish to determine the absorbed dose. In this case we can still employ the considerations leading to either Eq. 7-2 or 7-3. However, it is necessary to ensure that the ratio, D_x/D_W is nearly constant over the range of primary energies to be encountered, where D_x is the absorbed dose in a material x when the measured dose in the wall is D_W. It is necessary to determine the magnitude of D_x/D_W either by direct measurement or by calculation.

7-4 MEASUREMENT OF X- AND GAMMA-RAY DOSE AND DOSE RATE

Measurements of gamma-ray dose and dose rate are made with ionization chambers designed according to the Bragg-Gray principle. In Fig. 7-2 we show a cross section of a chamber with graphite (carbon) walls

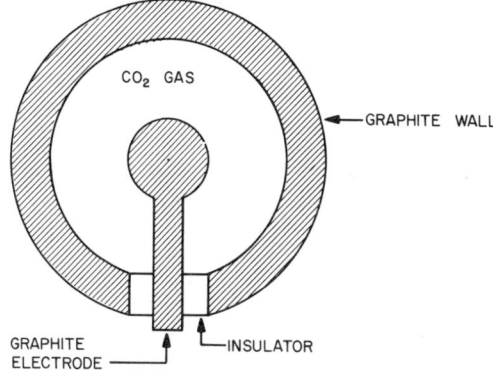

Fig. 7-2 Cross section of graphite-walled CO_2 ionization chamber.

and filled with CO_2 gas. We know from the Bragg-Gray principle that if the carbon wall has thickness greater than the range of secondary electrons, then the energy per gram of carbon can be determined by measuring the ionization in the CO_2 gas. It may be shown that, for photon energies between 0.2 MeV and 5 MeV,

$$\frac{D_t}{D_c} = 1.1(1 \pm 0.05) \tag{7-4}$$

where D_t is the energy absorbed per gram of soft tissue, and D_c is the energy absorbed per gram of carbon. Accordingly, the energy absorbed in tissue can be determined within an uncertainty of 5% by making measurements in a carbon chamber. Ratios of the kind appearing in Eq. 7-4 have been calculated for a number of substances, and some of the most useful ones are shown in Fig. 7-3. It is seen in this figure that if the photon energies are greater than 0.1 MeV, the absorbed dose ratios are nearly 1.0 for all substances composed of atoms having low atomic numbers. Therefore, many practical materials, such as plastics, are available for gamma-ray dose chamber (or dosimeter) walls.

The sensitivity of ionization chambers for the measurement of absorbed dose in tissue can be calculated by making use of Eq. 7-3 and assuming (1) that the ratio $D_w/D_t = 1$, and (2) that $S = 1$. Let us suppose that the chamber volume is 100 cm³ and is filled with air at one atmosphere pressure, so that the mass of the gas is about 0.13 g. Taking the W-value for electrons in air to be 34.0 eV per ip, we obtain $N_i = 2.4 \times 10^{11}$ ip, or 3.8×10^{-8} coul of positive and of negative charge per rad of energy absorbed. Suppose that a 4.0×10^{-9} farad condensor is charged to adequate potential so that when impressed on the chamber all ion pairs which are created are collected. The change in its potential due to one rad is then

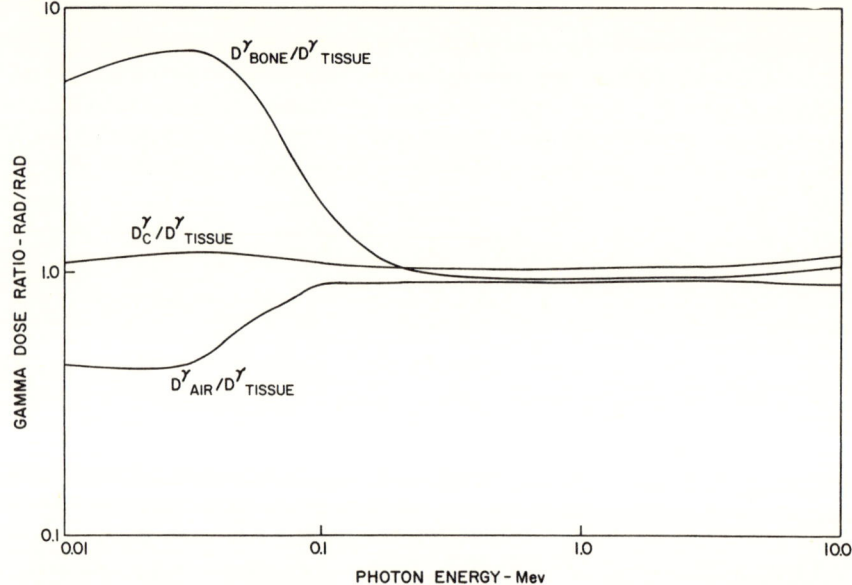

Fig. 7-3 Gamma-dose ratios for several substances as a function of photon energy.

about 10 V. These are the design considerations for condensor type ion chambers which are precharged to a measured potential, then exposed to gamma radiation. After exposure the new potential is determined and the change in potential is used to calculate the absorbed dose.

The measurement of dose rate is generally accomplished by measuring the current due to ionization. In our illustration above we saw that 1 rad produced 2.4×10^{11} ip in a chamber containing 0.13 g of gas. The current which flows when the dose rate is 1 rad/hr is about 10^{-11} amp. Simple electrometer circuits can be used to measure currents as small as 10^{-14} amp, and thus dose rates as low as 10^{-3} rad/hr.

7-5 MEASUREMENT OF NEUTRON DOSE AND DOSE RATE

Ionization chambers of the type described above will respond to neutrons having energies between about 0.1 MeV and 20 MeV. However, the response of the carbon wall-CO_2 gas ionization chamber, for example, varies considerably with neutron energy. One may calculate the neutron response $P(E)$ defined as

$$P(E) = D_c^n / D_c^\gamma \tag{7-5}$$

where $D_C{}^n$ is the energy absorbed in 1 g of carbon due to 1 *"tissue rad"* of fast neutrons and $D_C{}^\gamma$ is the energy absorbed in 1 g of carbon due to 1 tissue rad of gamma rays. The quantity $P(E)$ is approximately the response of a carbon-CO_2 chamber to fast neutrons, and its variation with neutron energy is shown in Table 7-1. We see from the table that large errors would be made in attempting to measure neutron dose with a carbon-CO_2 chamber, unless good knowledge is available of the neutron spectrum.

The obvious way to avoid the above difficulty is to build an ionization chamber in accordance with the Bragg-Gray principle so successfully used for X- or gamma-rays, but taking care that the atomic composition of both the wall and the enclosed gas approximate that of soft tissue, or other materials if the neutron dose in some other medium is sought. With respect to matching the atomic composition, the percentage of hydrogen is of great importance inasmuch as about 80% of the energy absorbed in tissue from fast neutrons is due to recoils with hydrogen atoms.

Successful *tissue-equivalent* chambers (Fig. 7-4) have been developed by Failla and Rossi of the Columbia University dosimetry group. Instruments made in this way can be used to measure the total (neutron plus gamma ray) dose or dose rate over a wide range of energies. Generally, however, one must be concerned with separate measurements of neutrons and gamma rays, because the biological effectiveness is quite different for the two radiations. For this reason two chambers are exposed to the mixed radiation field, one tissue equivalent, and one carbon-CO_2. The tissue equivalent chamber readings, R_T, correspond to the sum $(D_\gamma + D_n)$ and the carbon-CO_2 chamber readings, R_C, correspond approximately to $(D_\gamma + P(E)D_n)$. If $P(E)$ were equal to zero, R_C would give D_γ, and $R_T - R_C$ would give D_n. Information in Table 7-1 can be used to give $P(E)$

TABLE 7-1 Variation of P with Neutron Energy E

E(MeV)	P(%)
0.1	10.9
0.5	14.9
1.0	14.9
2.0	14.5
3.0	15.1
4.0	24.7
5.0	16.8
10.0	34.1
20.0	48.7

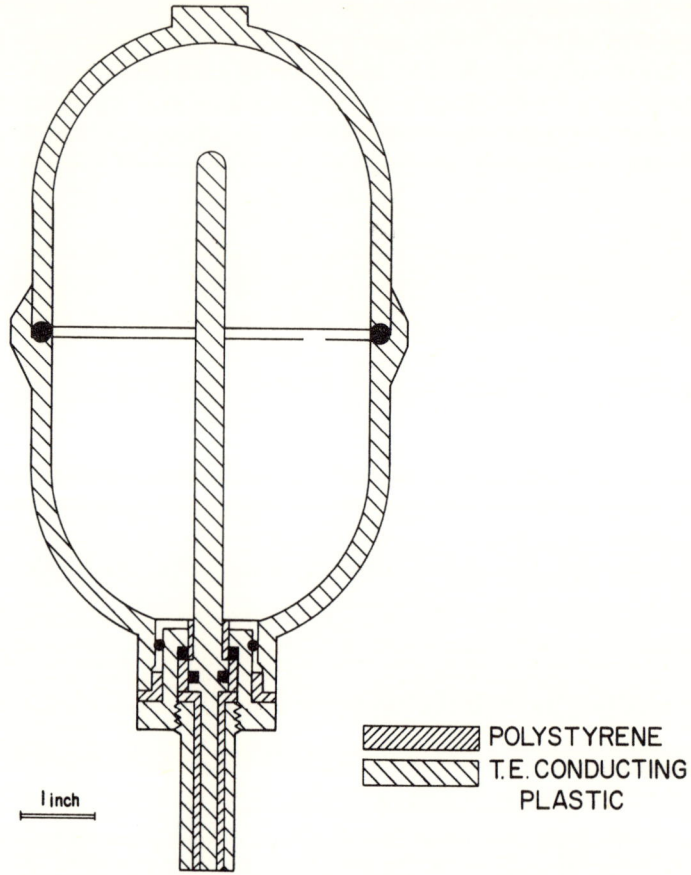

Fig. 7-4 Cross section of tissue-equivalent ionization chamber.

only when the neutron energy is well known, therefore the difference method to obtain D_n and D_γ is frequently not very accurate.

A more direct method of determining the dose due to fast neutrons makes use of a proportional counter, in which the height of the pulse is nearly proportional to the energy lost in the sensitive volume (Fig. 7-5). Consider the interaction of gamma rays and fast neutrons having comparable energies (e.g., 1 MeV). The 1 MeV gamma ray ejects an electron while the fast neutron typically creates a proton recoil. When the electron and proton have comparable energy, the electron has a very much longer range and a very much smaller dE/dX than the proton. Thus, if the cavity dimension is much smaller than the electron range, the energy deposited in the cavity will be much greater in the case of the proton. Pulse height analysis can, therefore, be performed electronically so that an instrument would register only recoil protons. These ideas have been used to develop proportional counter dosimeters which are very sensitive to fast neutrons and have a low response to gamma rays.

The proportional counter in Fig. 7-5, used to measure the tissue dose due to fast neutrons, follows the Bragg-Gray dosimetry principle. The wall is made of polyethylene (C_2H_4) plastic and the gas is either ethylene (C_2H_4) or cyclopropane (C_3H_6) so that the ratio H/C is the same for gas and wall in either case. The wall is made thicker than the ranges of a 20 MeV proton, and the gas pressure and cavity dimensions are selected to achieve good gamma ray rejection. To apply the Bragg-Gray principle, a measurement of the energy absorbed in the gas is needed. Pulse heights are, therefore, integrated; for example a pulse of 20 V at the output of an amplifier, is given a weight of 2 when a pulse of 10 V is given a weight of 1. Rejection of small pulses due to gamma rays, and subsequent integration of the larger neutron pulses, are done automatically with electronic binary scaling units. In this scheme the energy absorbed per gram of C_2H_4, D_E, is measured and, from this, the tissue dose is obtained by use of the equation $D_E/D_T = 1.45(1 \pm 0.10)$, which holds true in the neutron energy range between 0.01 MeV and 20 MeV. Fast neutron dose rates as low as 0.001 rad/hr can be measured even when the gamma ray background is as large as 100 rad/hr.

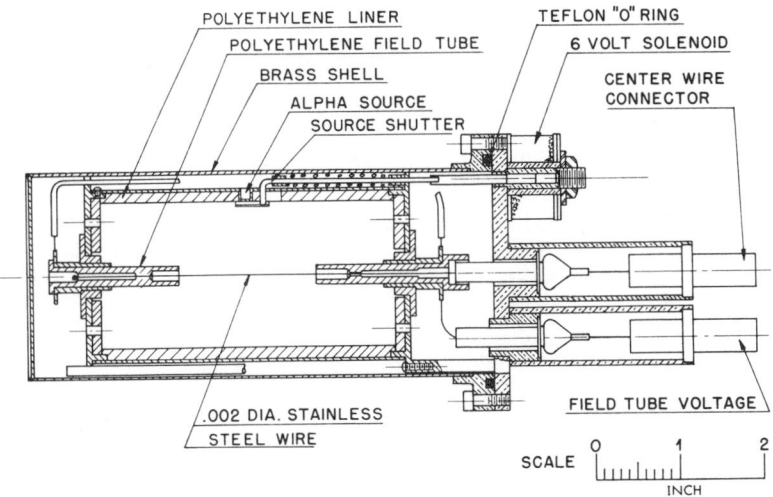

Fig. 7-5 Cross section of fast-neutron proportional counter.

90 Radiation Dosimetry

7-6 MEASUREMENT OF THE LET DISTRIBUTION

While the division of dose into light particles (e.g., electrons from gamma rays) and heavy particles (e.g., recoiling atoms from fast neutrons) is sufficient for most radiation protection standards, there are cases, for example, in biological experimentation, where it is desirable to measure the complete LET distribution. Proportional counters have been developed for this purpose by the Columbia University dosimetry group. Essentially, these devices measure the distribution of pulse heights due to charged particles traversing a cylindrical or spherical region. From these distributions the LET distribution is obtained by suitable data analysis.

7-7 PENETRATION OF NEUTRONS IN MATTER

As explained in Chapter 5 the penetration of neutrons in matter is a complex process, due to the fact that when fast neutrons scatter, neutrons of less energy are left over. Neutrons scatter repeatedly until finally they are thermalized and captured by an atom or escape entirely from the medium of consideration. Thus, it is impossible to discuss neutron penetration by means of the concept of range. Neither is it possible to define an absorption coefficient of the type that facilitated the discussion of X-ray and gamma ray penetration.

W. S. Snyder and J. Neufeld have carried out a very successful treatment of neutron penetration in tissue by making use of "Monte Carlo" calculations, in which games of chance are played with neutrons and atoms. In this game many thousands of neutrons are imagined to impinge on a medium whose atomic composition is specified, and all interactions of neutrons with atoms are guessed in an unbiased fashion by making use of tables of random numbers. Naturally, interaction cross sections are included as a rule of the game in order that the results are realistic. Further use of random numbers determines the direction and energy of the scattered neutrons and all subsequent interactions are followed and tabulated. After the neutron histories are determined and stored in the memory of electronic computers, the data may be processed to determine the locally absorbed dose resulting from charged particles produced by the interactions.

Figure 7-6 shows the absorbed dose in rad per incident neutron fluence (normal to the surface) for thermal neutrons in tissue. In this case the energy is deposited by two types of primary interactions. The interaction

$$_0^1 n + {}_7^{14}N \rightarrow {}_6^{14}C + {}_1^1 H$$

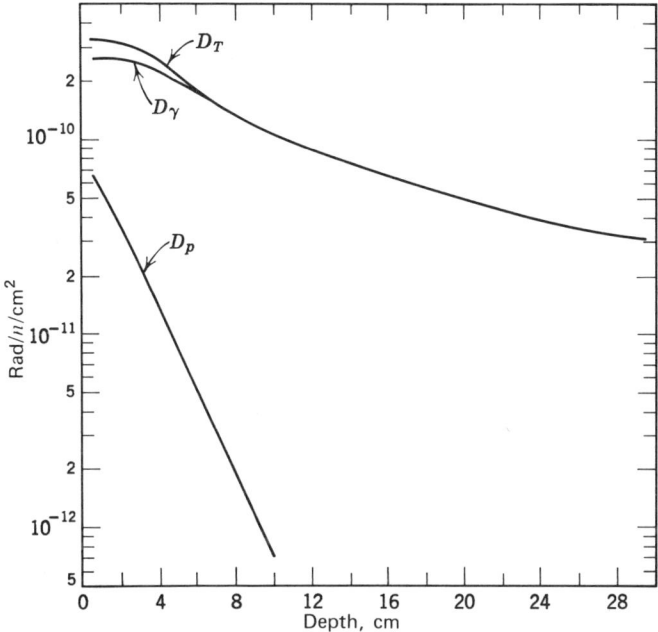

Fig. 7-6 Depth-dose curves showing the penetration of thermal neutrons in tissue. D_P, proton rad dose; D_γ, gamma rad dose; D_T, total rad dose. (From *Protection against Neutron Radiation up to 30 Million Electron Volts*, National Bureau of Standards Handbook 63, U.S. Department of Commerce, Washington, D.C., 1957).

gives rise to the curve labeled D_P and the interaction

$$_0^1n + {_1^1}H \rightarrow {_1^2}H + \gamma$$

gives rise to the curve labeled D_γ. From the total curve D_T, we see that thermal neutrons can cause the absorption of energy at great distances in tissue. Data on fast neutron penetration in tissue is presented in Fig. 7-7 for 1 MeV neutrons, and in Fig. 7-8 for 10 MeV neutrons. In these figures the component of absorbed dose designated as D_h is due to neutron collisions with heavy elements (the C, N, O components of tissue). In the case of fast neutrons, much of the absorbed dose is due to elastic collisions with hydrogen atoms (see the curves labeled D_P in figures 7-7 and 7-8). Notice in Figs. 7-7 and 7-8 that, even though the incoming neutrons are fast, they still contribute dose due to gamma radiation, D_γ. These curves reflect the fact that thermal neutrons are formed by "slowing down" collisions with atoms (mostly H atoms) after which gamma radiation is produced (e.g., by the hydrogen capture gamma radiation).

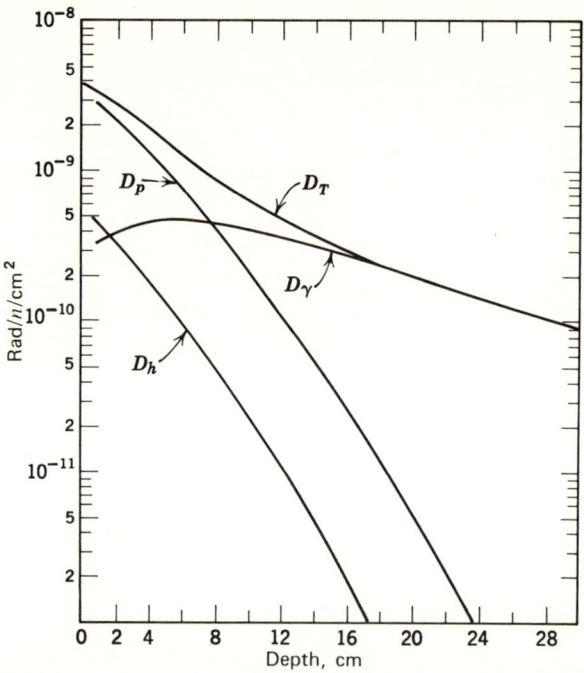

Fig. 7-7 Depth-dose curves showing the penetration of 1 MeV neutrons in tissue. D_P, protons; D_h, heavy recoils; D_γ, n–γ, reactions; D_T, total rad dose. (From *Protection against Neutron Radiation up to 30 Million Electron Volts*, National Bureau of Standards Handbook 63, U.S. Department of Commerce, Washington, D.C., 1957).

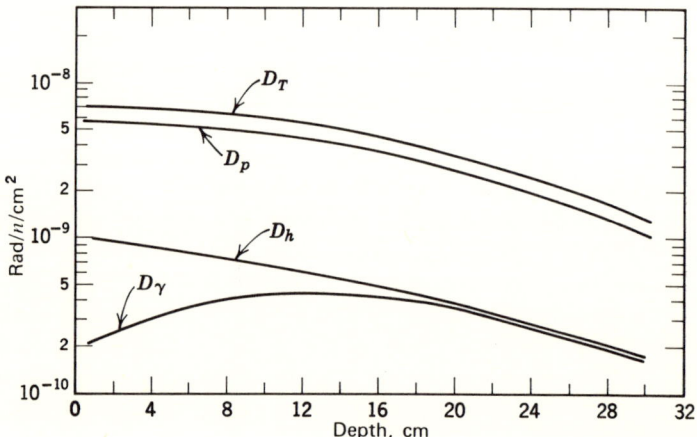

Fig. 7-8 Depth-dose curves showing the penetration of 10 MeV neutrons in tissue. D_P, protons; D_h, heavy recoils; D_γ, n–γ, reaction; D_T, total rad dose. (From *Protection against Neutron Radiation up to 30 Million Electron Volts*, National Bureau of Standards Handbook 63, U.S. Department of Commerce, Washington, D.C., 1957).

These gamma rays interact by Compton scattering and by the photoelectric effect, giving electrons with kinetic energy which is then absorbed in the medium to produce a dose. Curves of the type shown in Figs. 7-6 to 7-8 are known as depth-dose curves, and they show how the primary radiation is converted to absorbed dose as a function of depth in a medium.

PROBLEMS

1. An X-ray beam produces 10 esu of charge in 0.1 g of air per second. What is the exposure rate in R/sec? What current flows, if all of the ion pairs are collected?   [Ans. 0.078 R/sec, 3.3×10^{-9} amp]

2. A Van de Graaff generator delivers a 5 MeV proton beam current of 10^{-6} amp which is uniformly spread over a thin aluminum foil of 1 cm² area. What is the dose rate in units of aluminum rads sec^{-1}?   [Ans. 5.6×10^6 rad/sec]

3. Assuming that no secondary particles leave the aluminum foil in Problem 2, at what LET is the dose delivered?

4. If the stopping power for protons passing through a thin sample of aluminum is dE/dX and the density of aluminum is ρ (g cm^{-3}), derive a formula for the absorbed dose rate when the proton flux is ϕ (protons cm^{-2} sec^{-1}). (Hint. If dE/dX is MeV cm^{-1}, an acceptable unit for dose rate is MeV g^{-1} sec^{-1}.)

5. An ionization chamber designed according to the Bragg-Gray principle contains 0.1 g of CO_2 gas. A beam of gamma radiation creates a current of 10^{-10} amp under voltage saturation conditions. If W is 30 eV/ip what is the absorbed dose rate in CO_2 in units of rad/sec?   [Ans. 3×10^{-3} rad/sec]

6. An ionization chamber, designed according to the Bragg-Gray principle, has carbon walls and CO_2 gas. It is observed that, on exposure to a certain fluence of gamma radiation, 10^{12} ip are formed per g of CO_2. If the W-value of CO_2 is assumed to be 30 eV/ip, what is the absorbed dose in carbon in rad units?

7. Explain why the pulses generated in the proportional counter (Fig. 7-5) are smaller when a 1 MeV gamma ray produces a photoelectron than those generated when a 1 MeV neutron makes a head-on collision with a hydrogen atom.

8. Why is it important that a neutron dosimeter contain appreciable hydrogen, if an interpretation of absorbed dose in tissue is to be made?

9. Make a table showing the ratio of the rad dose due to thermal neutron interactions producing protons to the rad dose from gamma rays produced by thermal neutron interactions as a function of depth in tissue. Use 2 cm intervals for the depths. Explain why this ratio has a maximum near the surface.

10. Make a table showing the ratio D_P/D_γ, where D_P is the component of dose due to protons and D_γ is the component of dose due to gamma radiation, for

the depths 0, 2, 4, 8, and 10 cm in tissue when 1 MeV neutrons are incident to tissue. Explain why D_P/D_γ is much larger for 1 MeV neutrons than for thermal neutrons.

11. Repeat the table suggested in Problem 10 for 10 MeV neutrons. Why is the ratio D_P/D_γ always larger for 10 MeV neutrons than that for 1 MeV neutrons?

12. Calculate the 10 MeV neutron fluence which is normally incident on a slab of tissue when the total rad dose at 20 cm depth is 100 rad.

13. If a tissue-equivalent ion chamber contains 0.1 g of tissue-equivalent gas, what would be the ionization current if it were placed at a depth of 10 cm in a slab of water, when 10^6 neutrons cm^{-2} sec^{-1} are incident to the slab? Let the neutron energy be 1 MeV and the W-value of the gas be 34 eV/ip. What fraction of the current is due to gamma radiation? [*Ans.* 1.6×10^{-11} amp, 0.66]

14. Neglecting the neutron response of a C-CO_2 chamber, what ionization current would be read when a carbon-CO_2 chamber replaces the tissue-equivalent chamber in Problem 13, assuming that the chamber contains 0.05 g of CO_2 and the W-value of CO_2 is 36 eV/ip.

CHAPTER EIGHT

Biological Effects and the Control of Radiation Exposure

We shall take it for granted that the biological effects of radiations are due in some way to the chemical changes induced by the radiations. We are immediately faced with the problem of explaining why marked biological effects are produced by doses of radiation which produce only a small degree of chemical change.*

D. E. Lea

8-1 RELATIVE BIOLOGICAL EFFECTIVENESS (RBE)

The interaction of radiation with matter is accompanied by the absorption of energy. It is reasonable to expect — and experiments show — that the extent of biological damage produced in an organism as a result of irradiation depends on the amount of energy absorbed per gram (the absorbed dose) in the tissue of the organism. Since, as described in the last chapter, absorbed energy is relatively easy to determine in practice, the absorbed

---

*D. E. Lea in *Actions of Radiations on Living Cells*, Cambridge University Press, 1962. Quoted with permission.

dose is a convenient parameter upon which to base comparisons of different radiobiological studies. Experiment shows, however, that biological effects also depend on the spatial distribution of the energy deposited in tissue. In particular, we shall see the importance of the stopping power of a charged particle, or its linear energy transfer (LET)[1], in producing biological damage. Biological effects are also found to depend on the time distribution with which the absorbed dose is delivered.

The relative biological effectiveness of two radiations, RBE_1 and RBE_2, is defined in terms of the absorbed doses, D_1 and D_2, that produce the same biological effect under identical experimental conditions, by writing

$$\frac{RBE_1}{RBE_2} = \frac{D_2}{D_1}. \tag{8-1}$$

This definition does not imply a quantitative relationship between absorbed dose and biological damage. It merely defines the relative effectiveness in an experiment with two radiations that are made biologically equivalent by producing a specified amount of damage. Usually X-radiation is understood as the norm for specifying RBE.

8-2 LET EFFECTS

The values of RBE found in practice vary with experimental conditions, such as the energy and dose rate of the radiation, and the particular biological effect, or end-point, under study. A relationship exists between biological effectiveness and the LET of an incident charged particle. When the incident radiation is a neutron or photon, the LET of the secondary charged particles it produces is correlated with biological effectiveness. In this section we discuss a few experiments that show the effects of different LET on different types of biological organisms.

Figure 8-1 shows a comparison of RBE between relatively hard gamma rays, soft X-rays, and densely ionizing alpha particles. The horizontal axis gives the LET of secondary electrons produced by the photons and the LET of the alpha particles. LET is expressed here in units of kiloelectron volts (1 keV = 10^3 eV) per micron ($1\mu = 10^{-6}$ m) of path traveled in water. Compared with gamma rays, X-rays produce lower-energy secondary electrons having somewhat higher values of LET (Fig. 3-3). The data show an increase in RBE by a factor of about

---

[1] We regard LET and stopping power as synonymous in this discussion. In radiobiology, however, we distinguish between the energy lost by a charged particle and the energy actually absorbed in a volume of interest.

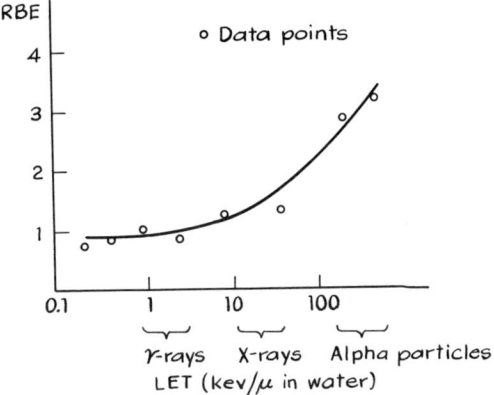

Fig. 8-1 RBE of different radiations for inhibition of cell division in yeast. (Based on data summarized by Raymond E. Zirkle, "The Radiobiological Importance of Linear Energy Transfer", in *Radiation Biology*, A. Hollander, ed., 1, p. 315, McGraw-Hill, 1954).

three in going from gamma rays to alpha particles, which have a much higher LET. Alpha particles inhibit cell division at much lower doses.

The results of another series of RBE experiments are shown in Fig. 8-2. Mice were exposed to acute doses from monoenergetic neutrons at several energies and a number of biological end-points were studied. As in the previous figure, we can interpret the results in terms of LET. Most of the neutron energy (e.g., at 10 MeV, 85%) is deposited in soft tissue through elastic collisions with hydrogen nuclei (protons). As shown in Section 5-3, the average proton recoil energy E_p in this energy range is one-half the incident neutron energy E_n: $E_p = \frac{1}{2}E_n$. Lower-energy neutrons, therefore, on the average produce lower-energy protons, having a higher LET (Fig. 3-4). As seen from Fig. 8-2, higher LET is more efficient in producing the effects studied. The absorbed dose required from the fastest neutrons is about four times greater than that from the lowest energy neutrons. As a rule, RBE values of ionizing radiation fall in the range 1 to 10.

The variation of RBE with LET does not appear to follow a universal rule, although several types of variation can be described. Curve A in Fig. 8-3 shows the type of behavior often exhibited, for example, by mamalian cells. As in the previous two figures, RBE increases with increasing LET, but only up to a point. Evidently, some kinds of biological damage are produced most efficiently (i.e., with the lowest absorbed dose) when the average rate of energy loss along a particle track has a certain value. Additional energy deposited at higher LET is wasted ("over-kill"), and the efficiency of the radiation is less. Curve B, in which the RBE is

98 Biological Effects and the Control of Radiation Exposure

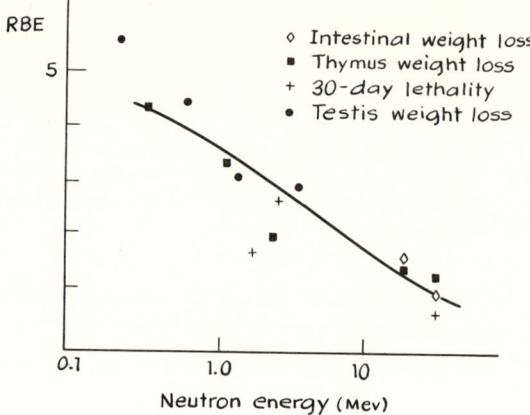

Fig. 8-2 Neutron RBE data for a number of biological endpoints produced in mice from acute exposure to monoenergetic neutrons. [Based on J. L. Bateman and V. P. Bond, "The Effects of Radiations of Different LET on Early Responses in the Mammal", *Annals of the New York Academy of Sciences*, **114**, 32 (1964)].

never greater than unity, is typical of some effects (for example, the inactivation of virus and enzyme molecules) produced in small biological objects. Because of its small size, the response of the object per unit absorbed dose is not increased by increasing the number of energy-loss events occurring along a particle track.

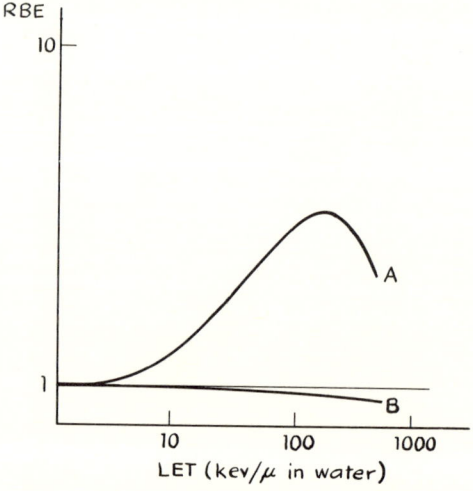

Fig. 8-3 Examples of RBE variation with LET.

8-3 DOSE-RATE EFFECTS

The dependence of RBE on the time distribution of absorbed dose—the dose rate—has been demonstrated in a number of experiments. We give here two examples.

The data points in Fig. 8-4 show the variation with dose rate of the LD-50—the absorbed dose required to produce 50 per cent lethality in 30 days—in a population of mice. At rates above approximately 250 rad/hour, the LD-50 is about 800 rad. At the lower dose rates the LD-50 is larger (about 1600 rad when the rate is 6 rad/hr).

Genetic studies with mice and *Drosophila* have shown that fewer mutations per rad occur when the dose rate is low than when it is high. Figure 8-5 shows experimental data for the mouse.

In producing other effects, a given dose can be less effective when delivered at once than when administered at intervals. The response of an organism appears to depend in a complicated way on many factors. A number of theories of damage and recovery have been proposed to account for observed dose-rate effects.

8-4 CELLULAR EFFECTS

Radiation produces immediate physical changes in a cell and its environment and may alter biological behavior at once or at some later

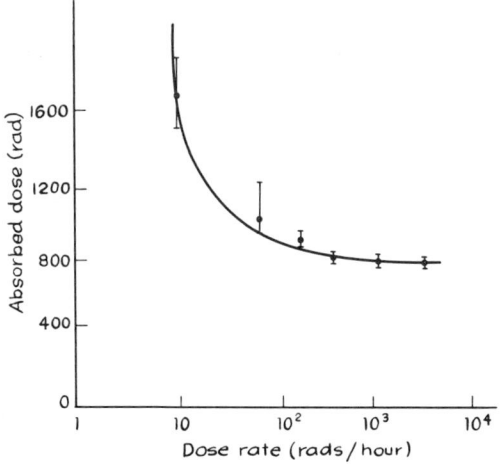

Fig. 8-4 Effect of dose rate on LD-50 for mice with ^{60}Co gamma radiation. [Based on J. F. Thomson and W. W. Tourtellotte, "The Effect of Dose Rate on the LD-50 of Mice Exposed to Gamma Radiation from ^{60}Co Sources", American Journal of Roentgenology, Radium Therapy, and Nuclear Medicine, **69**, 826 (1953)].

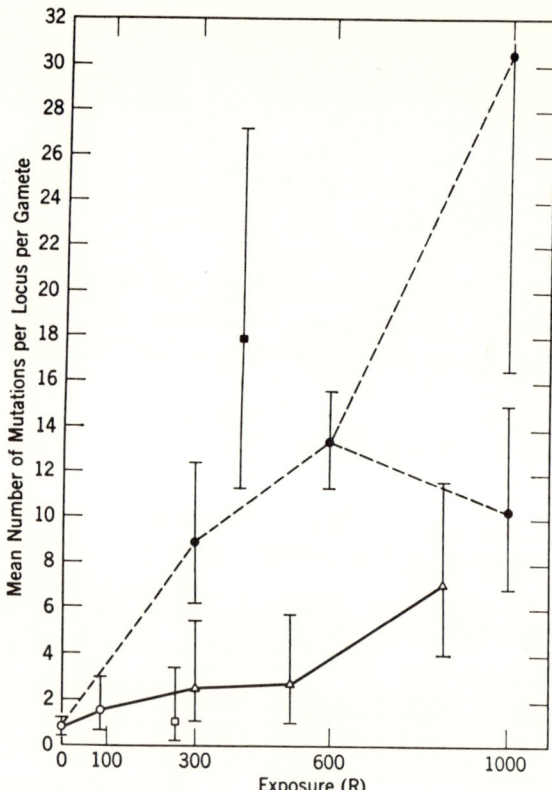

Fig. 8-5 Dose curves for specific locus mutations in the mouse. Ninety per cent confidence intervals shown. Solid points represent results with acute X-rays (80 to 90 R/min). Open points are chronic gamma-ray results (triangles and square, 90 R/week; circle, 10 R/week). Squares are mutation rates in females; all other points, mutation rates in males. The point for zero dose is the sum of all male controls. The top 1000 R point represents results of a single exposure; the lower one, results of successive exposures to 600 and 400 R, respectively, separated by an interval of 15 weeks. (Based on W. L. Russell, L. B. Russell and E. M. Kelley, in *Symposium on the Immediate and Low Level Effects of Ionizing Radiations*, Taylor and Francis, Ltd., London, 1960).

time. Specific cellular changes following irradiation include the breakup of molecules, the production of chemically active free radicals (atoms or groups of atoms with unpaired electrons), enzyme inactivation, alteration of deoxyribonucleic acid (DNA) synthesis, and chromosome breaking (Fig. 8-6). Gross physical changes, such as increased viscosity of cellular fluids, increased permeability of cellular membranes, swelling, and death can occur. When enough cells are damaged, the normal func-

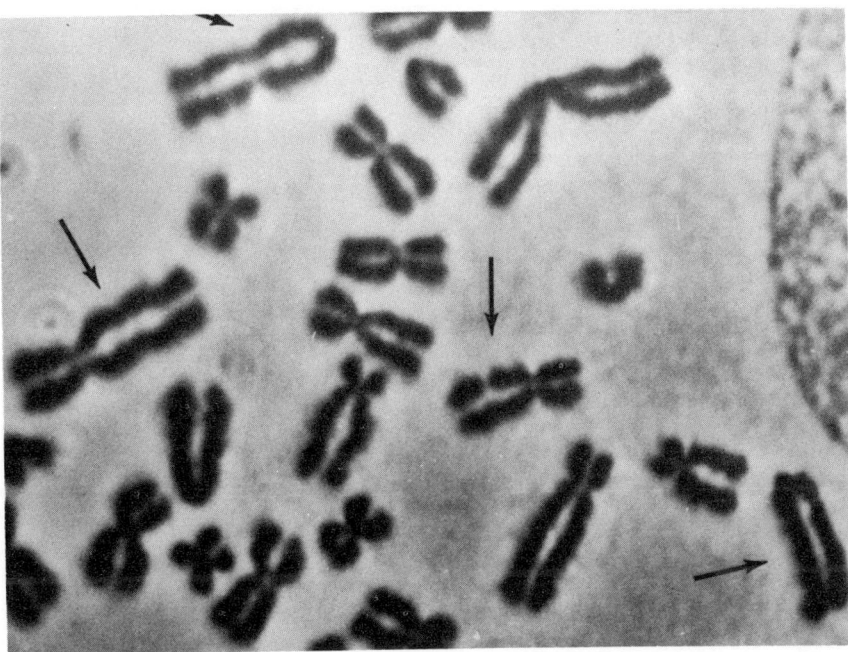

Fig. 8-6 Photomicrograph shows a metaphase spread of a pig kidney cell from a clone eight months after a dose of 500 rad (Argonne National Laboratory). Types of chromosome aberrations and recombinations, which persist and duplicate over many generations, can be seen (arrows). (Photo courtesy U.S. Atomic Energy Commission.)

tioning of an organ or of an entire organism can be impaired. Although many aspects of radiation injury are known from experiment and clinical observation, the detailed mechanisms and steps that lead from the absorption of radiation to the appearance of symptoms are not known. The subject of the production of radiation damage in biological systems is under intensive research and study.

Different kinds of cells and organs show different sensitivity to radiation. As a general rule, rapidly dividing cells, such as those often occurring in a malignant growth, are least resistant. The most radiosensitive cells in the body are those of the bone marrow, lymphoid, and epithelial tissues. As described in Section 8-6, blood changes, resulting from bone marrow damage, and the destruction of lymphocytes, are among the earliest observed effects following the absorption of radiation. At low doses, microscopic changes in the blood may be the only evidence of irradiation. Damage to epithelial cells produces injury in the gastrointestinal tract and in the reproductive organs. Epithelial cell damage may also result in epilation, or loss of hair. Cells of the fetus are also especially

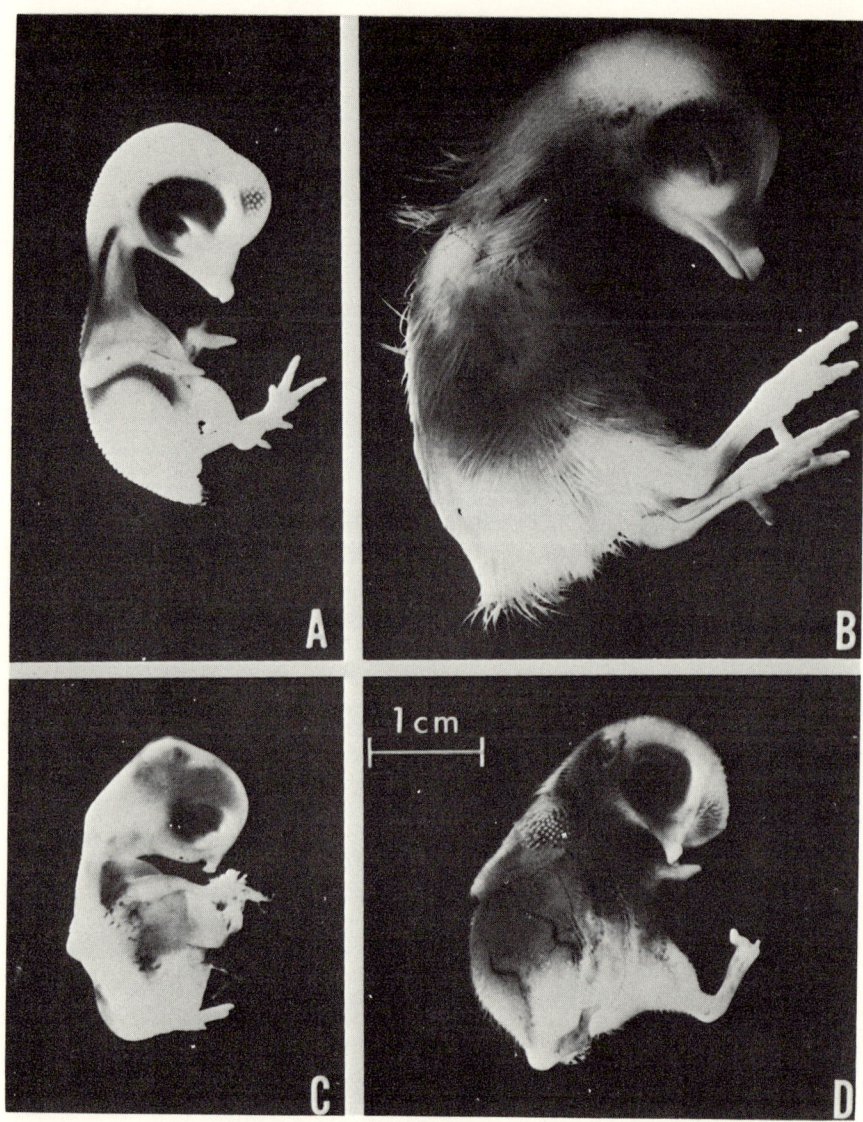

Fig. 8-7 Effects of ionizing radiation on the chick embryo. (*A*) 10 day non-radiated embryo. (*B*) 13 day non-radiated embryo. (*C*) 10 day embryo which had been exposed to ^{60}Co gamma radiation 4 days earlier. Note deformities of the beak and toes and generalized edema and hemorrhages. (*D*) 13 day embryo following exposure to ^{60}Co gamma radiation as in (*C*) except that radiation was applied 7 days before this picture was taken. In addition to the effects seen in (*C*) there is serious growth retardation. (Photo courtesy Argonne National Laboratory).

sensitive to radiation (Fig. 8-7). Less radiosensitive are cells of the bone, muscle, and blood vessels. Nerve cells show the most resistance to radiation damage.

8-5 THE QUESTION OF THRESHOLD

At present it is not known whether an absorbed dose greater than a certain minimum, or threshold, amount is necessary for the production of biological damage from radiation. Although the effects just described are demonstrable with large amounts of radiation, doses can be made so low that no visible effects occur. Figure 8-8, as an example, shows how the probability per year (incidence) of leukemia in man varies with radiation dose. Studies of Japanese survivors from the atomic bombs at Hiroshima and Nagasaki and of patients receiving theraputic X-ray treatments indicate that the solid curve describes the relationship between dose and incidence when the dose is several hundred rad or greater. Below about 100 rad too few cases of leukemia occur in the populations studied to demonstrate clearly an increased incidence due to radiation.

At doses of 10 rad or 1 rad, based on the observed incidence at high doses, the size of an exposed population needed to show a significant statistical increase would be enormous. Experiments with animal populations are not feasible because of the large number of individuals required. We must hypothesize what effect low doses might have on leukemia incidence. One assumption frequently made is that the approximately

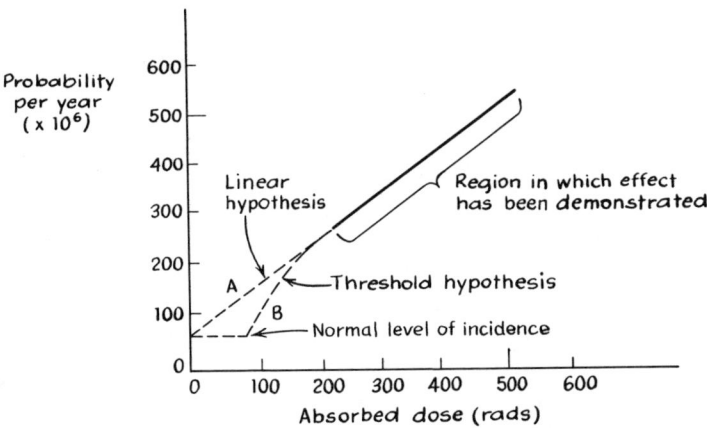

Fig. 8-8 Incidence of radiation induced leukemia in humans. Curve A illustrates a linear hypothesis for the effect of low doses and curve B, a threshold hypothesis.

104   Biological Effects and the Control of Radiation Exposure

linear solid curve at high doses should be extrapolated to zero dose, as shown by the dashed curve A in Fig. 8-8. According to this hypothesis, any amount of radiation, no matter how small, increases the probability of leukemia above its normal value by an amount proportional to the dose. Another assumption is that the body can repair damage done by small amounts of radiation. According to this hypothesis, the body tolerates low doses, and a certain threshold dose is required before the probability of leukemia is increased above its normal value. This threshold hypothesis is illustrated by the dashed curve B in Fig. 8-8.

Curves with shapes other than those of A and B have also been postulated for connecting the high dose region, in which data are statistically significant, to the region of low dose. Differentiating between a linear or a threshold response for any biological effect is complicated by the fact that radiation-induced effects occur with a natural (i.e., nonradiation induced) incidence.

8-6   THE ACUTE RADIATION SYNDROME

The characteristic symptoms shown by a person whose whole body, or a large portion of it, is exposed to a dose of radiation in a short time is called the acute radiation syndrome. These symptoms have been studied in animal experiments and, more importantly, in cases of radiation administered to persons in hospitals, in the atomic bomb survivors at Hiroshima and Nagasaki, and in persons exposed accidentally to large doses of radiation. To be specific, we discuss the acute radiation syndrome for gamma radiation, which, because of its penetrating power, gives a more or less uniform dose throughout the body. We must keep in mind that biological response to radiation varies from one individual to another, and so the numbers given here represent only an average expected among a number of exposed persons.

Table 8-1 summarizes the principal characteristics of the acute radiation syndrome. With doses below about 25 rad, little or no detectable effects are expected. With doses from 25 rad to 100 rad, detectable changes occur in the blood. Near 100 rad, the lymphocyte count decreases within a matter of hours after exposure to the radiation. After one or more days the white blood count falls and may subsequently fluctuate. The count of blood platelets, which are necessary for clotting, also begins to decline shortly after exposure. The change in the red blood cell picture is not as rapid and may not show a decrease for several days.

A dose of 100–300 rad can result in some radiation sickness, but probably will not, in itself, be fatal. Nausea and vomiting usually occur within

TABLE 8-1 Acute Radiation Syndrome

Gamma-Ray Whole-Body Dose (Rad)	Effects	Remarks
0–25	None detectable	
25–100	Blood changes; person feels little or no effect.	Lymph nodes and spleen damaged; lymphocyte count drops. Bone marrow damaged; decrease in white blood cell, platelet, and red blood cell count.
100–300	Blood changes, vomiting, malaise, fatigue, loss of appetite.	Antibiotic treatment may be necessary. Recovery can be expected.
300–600	Above effects plus hemorrhaging, infection, diarrhea, epilation, and temporary sterility.	Antibiotics and blood transfusions administered. Expect recovery in about 50% of cases at 500 rad. Possible bone marrow transplant.
More than 600	Above symptoms in addition to damage to the central nervous system; incapacitation at doses in excess of about 1000 rad.	Death almost a certainty. Sedation. Possible bone marrow transplantation in lower portion of this range.

a few hours in the range of 200 to 300 rad. Symptoms such as malaise, fatigue, and loss of appetite can be expected to appear. The blood picture will change dramatically in the manner already described, and many months may elapse before it returns to normal. Antibiotics may be required to offset the danger of infection due to the decline in the number of white blood cells (leukopenia). If no complications develop, recovery of the exposed individual can be expected.

Doses in excess of 300 rad may be fatal. In addition to changes already noted for lower doses, diarrhea and hemorrhaging can occur. The dangers of infection are compounded by lowered resistance of the body. The gastrointestinal tract is severely damaged: epithelial cell linings ulcerate, and uncontrolled bacterial infections may develop. Epilation (Fig. 8-9) may occur in about two weeks, and temporary sterility can be expected. Theraputic measures include the use of antibiotics, blood transfusions, and, at high dose levels, bone marrow transplantations.

Only a small percentage of individuals would probably survive an acute whole-body dose of 600 rad. Experience with humans in this high dose range is limited to a small number of persons exposed accidentally.

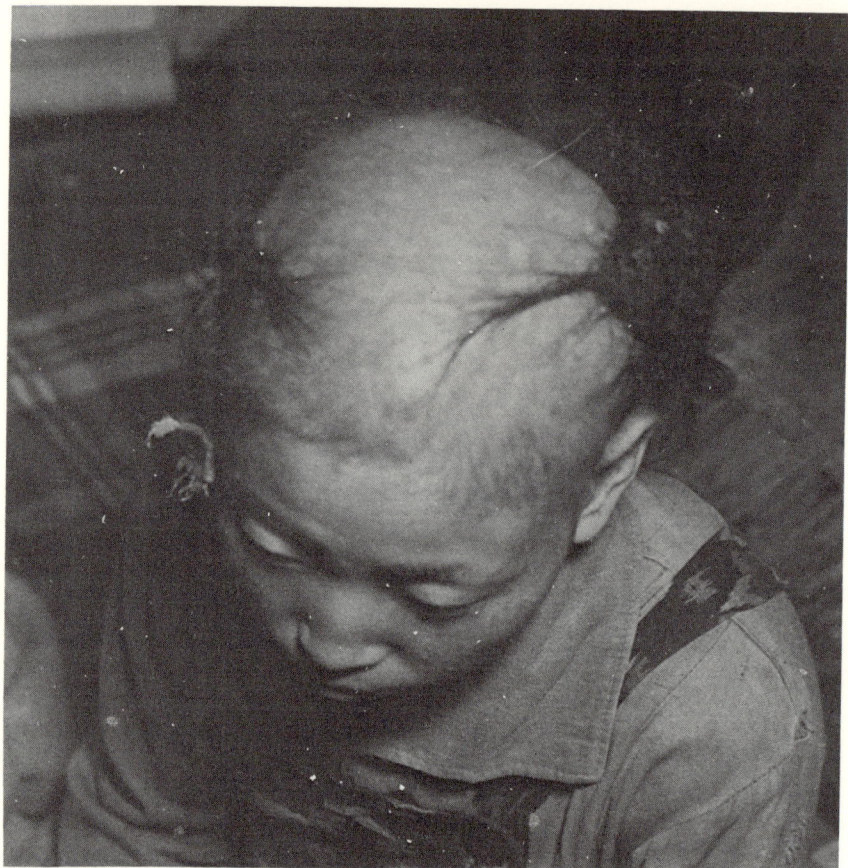

Fig. 8-9 An example of epilation due to radiation exposure. (Photo courtesy U.S. Atomic Energy Commission).

Doses up to several thousand rad cause death within one or two weeks. Larger doses result in death within days. In addition to affecting the gastrointestinal tract, high doses damage the central nervous system, and an exposed individual can be almost immediately incapacitated. Death can be caused by circulatory or respiratory failure.

8-7 DELAYED SOMATIC EFFECTS

Somatic effects of radiation are those seen or detected in the exposed individual, as contrasted with *genetic* effects (Section 8-8), which show up in his progeny. In addition to the acute somatic changes discussed in the last section, delayed effects are caused by radiation; the latent period between exposure and appearance of a symptom being months or years. Since the somatic and genetic effects produced by radiation also occur with normal spontaneous incidence in populations, it is often impossible to ascribe to radiation the occurrence of a symptom in a particular exposed individual. Delayed somatic effects due to radiation in a population occur statistically; an individual in that population is presumably placed at a higher risk of injury. In this section we describe several delayed effects known to be produced by radiation.

We observed in Fig. 8-8 the correlation between high radiation doses and the probability per year of the occurrence of leukemia. The peak incidence of the disease among the survivors at Hiroshima and Nagasaki after the bombings in 1945 occurred during the period 1950 to 1952, an abnormally high incidence continuing beyond this time. Further studies of these exposed populations showed evidence of an increased incidence of other types of neoplasia.

An example of diseases produced in humans, as a result of radiation exposure over a long period of time, includes bone cancers in the radium dial painters discussed at the end of Section 2-3. Like leukemia, such effects have also been demonstrated in laboratory experiments with animals.

Studies of irradiated animals also showed a shortening of the life span not attributable to any specific cause. Experiments with mice indicate that a life shortening of 1 or 2% occurs as a result of acute radiation doses of about 100 rad.

The production of radiation-induced lens opacities and cataracts has been demonstrated (Fig. 8-10). Acute exposure to radiation appears to be more serious than protracted exposure in producing cataracts, which may appear several years after the exposure. The X-ray dose required to produce clinically significant cataracts is probably at least 400 rad. Fast neutrons are much more effective. RBE values for cataract production in excess of 10 have been reported.

The irradiation of pregnant women can have adverse effects, some of which are manifested in subsequent years by the child. In addition to abnormally high still-birth and infant mortality rates among Japanese survivors, pregnant at the time of the atomic bombings, there appears to be an above-normal amount of mental retardation in the offspring. Small children, who were exposed, also appear to lag somewhat behind their contemporaries in overall physical development. These effects are noticeable only among survivors who received doses of several hundred rad. An example of a deformity in a calf from a cow, irradiated during gestation, is shown in Fig. 8-11.

108 Biological Effects and the Control of Radiation Exposure

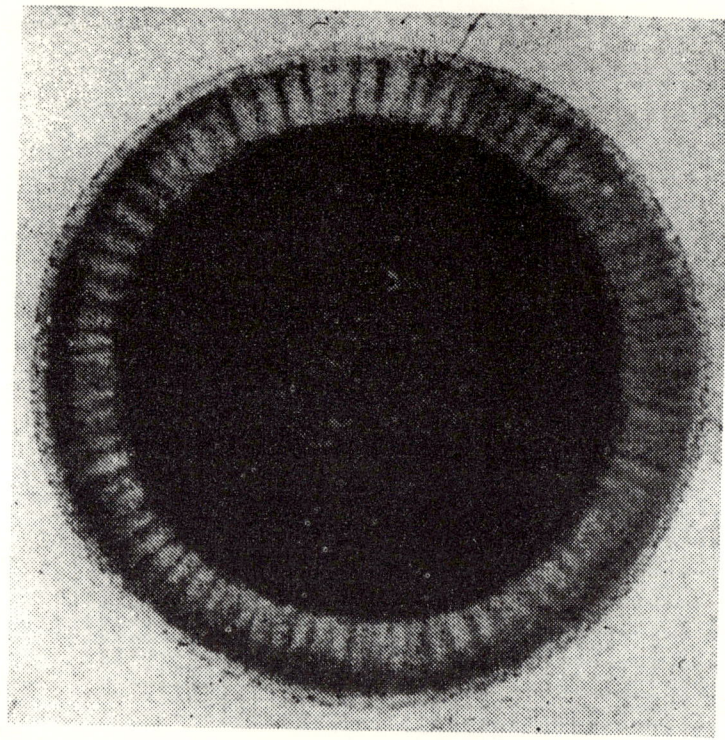

Fig. 8-10 Rabbit lens 255 days after acute exposure to 1500 R of X-rays. Slight opacities, consisting of thickened lens fibers, extend radially from a central opacity. [From F. Kandori, American Journal of Ophthalmology, **41**, 1006 (1956)].

Doses of a few thousand rad have been given with beta rays to the skin of the body without affecting deeper lying organs. In many cases a chronic abnormal condition of radiodermatitis developed over a period of years after exposure. The skin became thin and susceptible to ulceration and the formation of cancer. An example of beta-ray burns is shown in Fig. 8-12.

8-8 GENETIC EFFECTS

Gene mutations occur spontaneously in nature. The rate in a population can be increased by elevated temperatures, ultraviolet light, and certain chemical agents. Muller, in 1927, discovered that ionizing radiation also produces mutations.

Although mutations can be produced in any cell of an individual, only

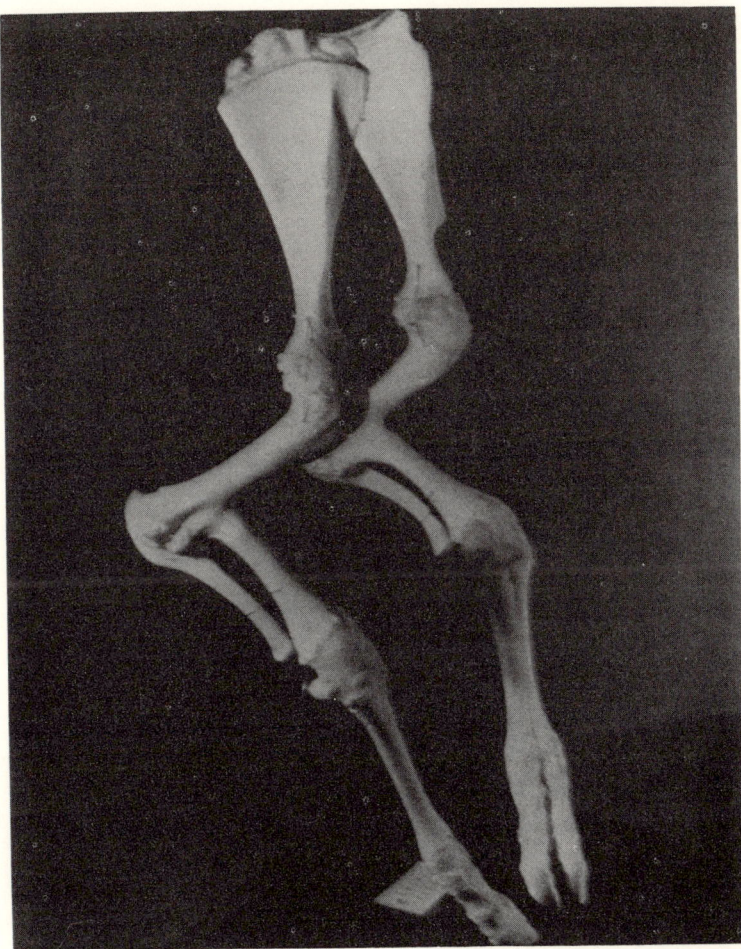

Fig. 8-11 Front legs of calf born to cow that received a whole-body gamma dose of 400 rad on the 32nd day of gestation (University of Tennesse—AEC Agricultural Research Laboratory). Other fetal calves irradiated two days later suffered only minor damage to the toes. (Photo courtesy U.S. Atomic Energy Commission).

those produced in a germ cell can be transmitted to his progeny. A mutant gene introduces changes that may or may not be apparent in the next generation. Genetic changes in an offspring may be minor or inconsequential, or they may be so serious that he is severely handicapped. Figure 8-13 illustrates some severe genetic changes found in the offspring of an irradiated fruit fly population.

The frequency of radiation-induced mutations depends on the absorbed

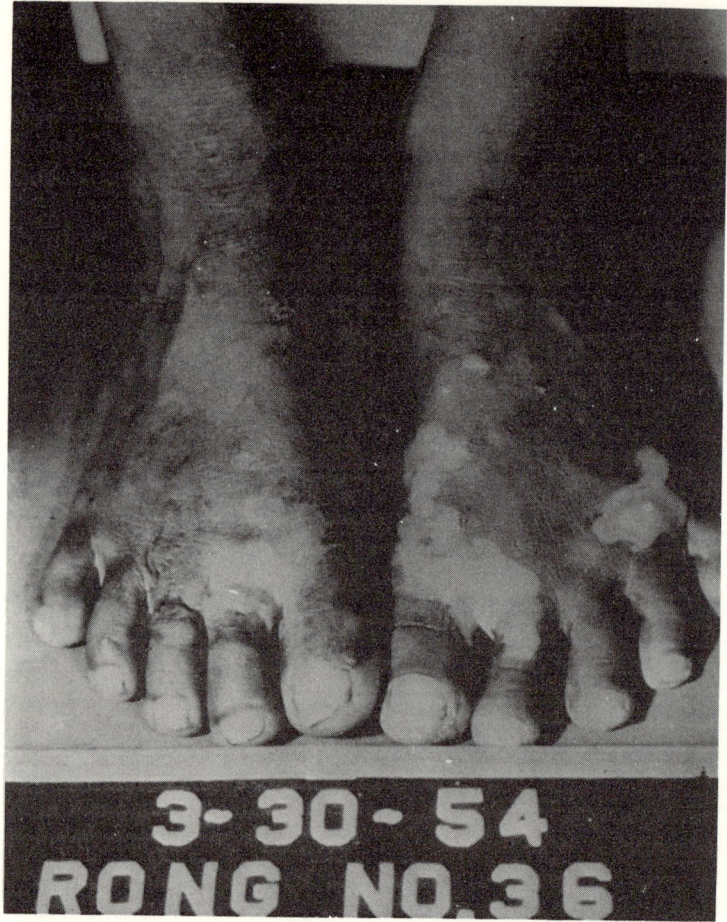

Fig. 8-12 Beta-ray burn to the feet (a) one month and (b) two months after exposure. Such exposures occurred accidently to a group of Marshallese natives, Japanese fishermen, and Americans from fallout from a hydrogen weapon tested at the

dose and the dose rate. An increase in mutation rate in a large experimental population becomes noticable with absorbed doses of a few tens of rads, and thereafter it appears to increase linearly with increasing dose. The effect of dose rate is illustrated by Fig. 8-5.

Both the frequency and the type of mutation also depend on the time of mating after exposure. When mating takes place soon thereafter, the germ cell involved has a high probability of having been mature at the time of irradiation. Irradiated mature germ cells carry a relatively high inci-

Genetic Effects 111

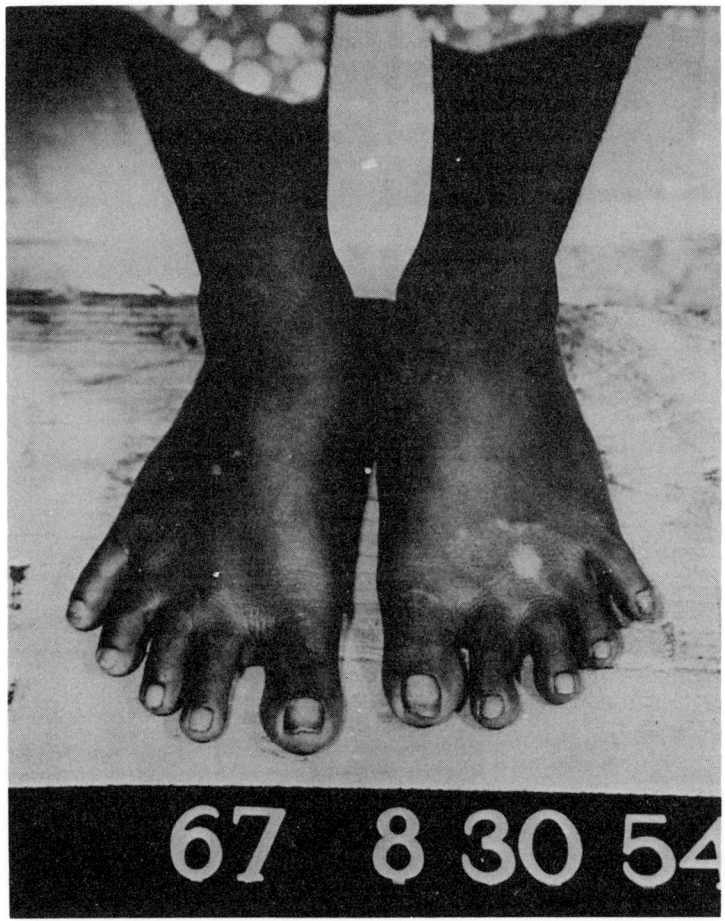

Bikini Atoll in the spring of 1954. Simple washing of the fallout material from the skin would have greatly reduced the radiation exposure. (Photo courtesy U.S. Atomic Energy Commission).

dence of dominant lethal mutations. When the mating occurs later, there is a relatively high probability that it will involve a cell that was immature when irradiated. Mutations arising in late matings following exposure are predominantly recessive.

Since its initial discovery, radiation genetics has become an active area of research (Section 10-6). Radiation studies furnish information not only about the effect of radiation on genes but also about the effects of genes on developmental processes.

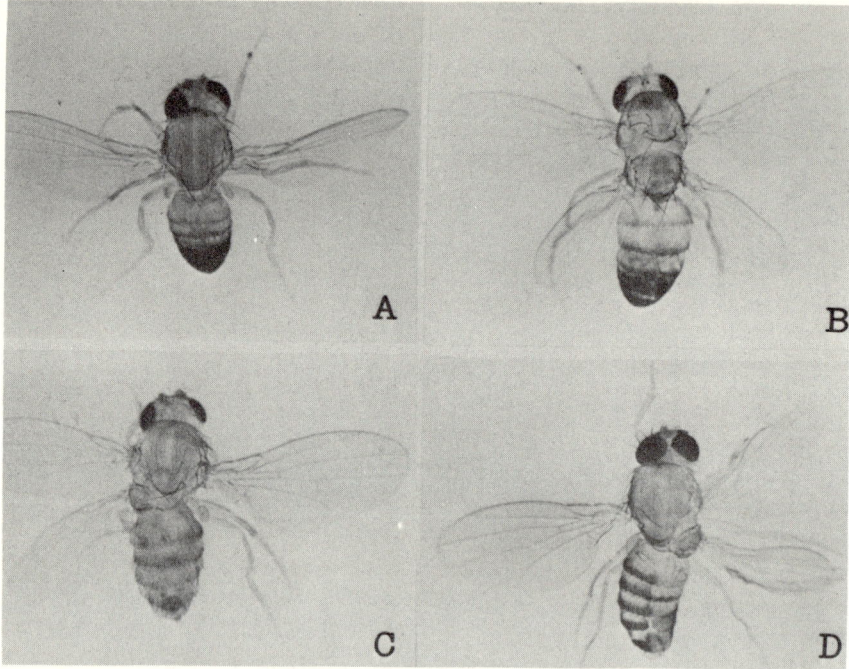

Fig. 8-13 Genetic radiation damage (California Institute of Technology). X-irradiation here produced fruit flies (*Drosophila melongaster*) with three and four wings and double or partially doubled thoraxes. Photos above show: (*A*) normal male Drosophila; (*B*) four-winged male which has double thorax; and (*C*) and (*D*) three-winged flies with partial double thoraxes. (Photo courtesy U.S. Atomic Energy Commission).

8-9 RADIATION PROTECTION STANDARDS

We have pointed out that a number of early investigators received burns from X-rays and radioactive materials. As the use of radiation became more widespread and its deleterious effects better recognized, the need arose to establish permissible levels of exposure to radiation for workers. In the year of its formation (1925), the International Commission on Radiological Units (ICRU) recommended that the annual permissible dose of radiation be one-tenth of the erythema dose — the acute dose that just produces visible reddening of the skin. The International Commission on Radiological Protection (ICRP) and, in the United States, the National Committee on Radiation Protection (NCRP) were also organized. Through the years, the recommendations of these three organizations have not differed widely. In the mid-1930's, the permissible exposure for whole-body X-radiation was set at 1 R/week by the ICRP and 0.5 R/week by the NCRP.

After World War II, routine operations in the atomic energy industry exposed workers to neutrons as well as radiation from radioactive materials. In setting standards for different kinds and energies of radiation, it is necessary to take account of their different biological effectiveness, as described in Section 8-1. To this end, the ICRU and ICRP in 1962 recommended that quality factors (QF) be used for different kinds of radiation.[1] They recommended that the values of QF given in Table 8-2, based on the linear energy transfer (LET) of the radiation, be used for long-term, occupational exposure to radiation.[2] To assess radiation exposure for control and protection purposes, the absorbed dose from radiation in different LET ranges is multiplied by the appropriate quality factor. The resultant number is then called the dose equivalent (DE), expressed as rem (roentgen-equivalent-man). When all of the absorbed dose D_a (rad) is from a single range of LET, the dose equivalent is given by

$$DE = QF \times D_a \qquad (8\text{-}2)$$
$$(\text{rem} = QF \times \text{rad}).$$

The QF value for X, gamma, and beta radiation is unity. For other radiations the QF's are generally two or three times larger than RBE values. The dose equivalents in different LET ranges are added when monitoring different kinds and energies of radiation exposure.

Maximum permissible levels of radiation in general use throughout

Table 8-2 LET-QF Relationship Recommended by ICRP and ICRU

LET (keV/micron in Water)	QF
3.5 or less	1
3.5–7.0	1–2
7.0–23	2–5
23–53	5–10
53–175	10–20
Gamma rays, X-rays, electrons and positrons of any LET	1

[1] "Radiation Quantities and Units," ICRU Report 10a, National Bureau of Standards Handbook 84, Washington, D.C. (1962).
[2] "Report of the RBE Committee to the ICRP and ICRU," published in *Health Physics*, **9**, 381 (1963).

the world for long-term occupational exposure are given in the 1965 report of the ICRP.[3] Some of these recommendations are shown in Table 8-3. The levels apply only to exposures received by an individual as a result of routine occupational work, apart from his exposure for medical or dental reasons. It is seen that the most radiosensitive organs are considered to be blood-forming organs, gonads, and lens of the eye, all of which are alloted an average 5 rem/yr. The maximum permissible dose equivalent for most other organs is 3 times this amount, while that for the thyroid is 6 times and that for the extremeties 15 times as great. The maximum permissible levels recommended for the public are generally one-tenth those for radiation workers. Exposure of the public occurs principally as a result of the escape from controlled radiation areas of small amounts of radioactive material in the air and in liquid effluents. The dose equivalent to the worldwide population is limited primarily for genetic reasons.

The Federal Radiation Council (FRC) was established in the United States in 1959 for the purpose of promulgating the radiation protection policy of the federal government in its operations that involve radiation. The Council, like the ICRP, recognizes that radiation protection standards should reflect a balancing of risk and benefit. The balancing cannot, of course, be carried out in a precise manner, since it is necessarily based on many social and moral judgements as well as scientific knowledge. The Council has recommended a set of "radiation protection guides" to be applied to operations connected with the peaceful uses of atomic energy. These guides are consistent with the maximum permissible levels in Table 8-3. Exposures entailed in the testing of nuclear weapons, for example, based on military and national defense requirements, have not been considered by the FRC.

In order to limit exposure from radioisotopes that can be inhaled or ingested, the ICRP and NCRP have recommended maximum permissible concentrations (MPC's) in air and water. MPC values have been recommended both for use in monitoring controlled areas and for monitoring the effluents from a controlled area into the environment. The values are estimated to be low enough to assure that 50-year exposure to concentrations not exceeding the MPC's will not give more than the maximum permissible rate of dose-equivalent to any organ. The estimates are based on normal eating, drinking, and breathing habits, and they take account of the different ways in which the body handles different chemical elements. As a rule, one organ in the body is the "critical organ;" the dose to this organ limits the total amount of a radioisotope that may be taken into the body. Exceeding the MPC termporarily does not necessarily result in an overdose since, as seen in Table 8-3, the per-

Table 8-3 Maximum Permissible Dose Equivalent Levels Recommended by the ICRP

Exposed Group	Blood-Forming Organs, Gonads, Lens of Eye	Other Organs
Radiation workers in controlled area	5 rem/yr average after age 18; no more than 2.5 rem in any 3 mo period	Skin, bone, thyroid: 30 rem/yr Hands, feet, forearms: 75 rem/yr Other organs: 15 rem/yr
Members of the public in the vicinity of a controlled area	0.5 rem/yr	Skin, bone, thyroid: 3.0 rem/yr for adults; 1.5 rem/yr for children up to 16 years of age Hands, feet, forearms: 7.5 rem/yr Other organs: 1.5 rem/yr
Population at large (worldwide)	5 rem average to age 30	

missible doses are averages over a length of time. The FRC introduced the concept of "radiation concentration guides" for use instead of MPC's. The guides specify what remedial action, if any, is to be taken when concentrations of radionuclides in the environment rise to certain stated levels.

Medical and dental X-rays which, as seen from the table given in the Foreword, constitute the greatest single source of exposure to the population of the United States, are not under the control of the federal government. The balancing of risks and benefits in this case is left to the medical judgment of the licensed physician. Some states have passed regulatory measures and provide surveys to see that medical X-ray equipment meets acceptable standards of operation. A number of state laws also apply to the use of X-ray machines and fluoroscopes for other purposes, for example, for shoe fitting and for luggage inspection. Many unfortunate cases of radition injury have resulted from the indiscriminate use of X-rays and radioactive isotopes not under the relatively new control and licensing procedures of the federal and state governments. In many instances, sources of radiation have been novel, and their hazards simply not recognized.

[3]"Recommendations of the International Commission on Radiological Protection" (Adopted September 17, 1965), ICRP Publication No. 9, Pergamon Press, London (1966).

PROBLEMS

1. The curve in Fig. 8-1 begins to rise in the region (1-10 keV/μ) where the average distance between ions is approximately 3×10^{-6} cm. Discuss the possible significance of this fact.
2. Based on LET, why should the value of RBE for alpha particles be greater than that for beta particles in producing certain biological effects? Give one or more hypotheses to explain why radiation of high LET should have a larger RBE than radiation of low LET.
3. Estimate from Fig. 8-2 the relative absorbed doses of 0.2 MeV and 10 MeV neutrons needed to produce the biological effects considered there. [*Ans.* 3]
4. Give one or more reasons why two different radiations, having the same LET in Fig. 8-3, might be expected to have a different RBE for producing certain effects in organisms of different size (e.g., in a virus and in a human cell).
5. Explain how the data in Figs. 8-4 and 8-5 might indicate that biological systems can repair radiation damage.
6. In routine operation in atomic energy plants in the United States, a maximum annual dose equivalent of 5 rem is allowed to the trunk of the body whereas a maximum of 75 rem is allowed to the hands and feet. Why is a greater dose equivalent allowed to the extremeties?
7. Because radiation kills ordinary healthy cells in the body, why is it nevertheless often used in the treatment of cancer?
8. Estimate from curve *B* in Fig. 8-8 the minimum dose needed to increase the risk of leukemia, based on the threshold hypothesis. [*Ans.* 80 rad]
9. Give at least one reason "for" and one "against" both the linear and threshold hypotheses of radiation damage.
10. Explain why experiments designed to differentiate between the linear and threshold (or other) hypotheses involve very large control and exposed populations.
11. In the acute radiation syndrome, how is hemorrhaging related to the drop in the blood platelet count?
12. What is meant by the expression "gastrointestinal death" from acute radiation exposure?
13. Can one conclude from the lack of detectable symptoms occuring below about 25 rad (Table 8-1) that doses of this magnitude and less are inconsequential? Discuss.
14. Explain the statement at the beginning of Section 8-7 that delayed somatic effects of radiation in a population are statistical in nature.
15. List at least four late somatic effects produced by radiation.
16. Are any late somatic effects uniquely caused by radiation?
17. Compare the values of stopping power (LET) for protons given in Fig. 3-4 with the quality factors in Table 8-2. What *QF* applies to a proton of energy 2 MeV in soft tissue? (Stopping powers of water and soft tissue are approximately the same.) [*Ans.* 2 to 5]

18. What is the dose equivalent from an absorbed dose of 20 rad of X-rays? What is the dose equivalent of 20 rad of neutrons with $QF = 8$?
[*Ans.* 20 rem, 160 rem]

19. What absorbed dose from protons with $QF = 3$ gives the same dose equivalent as 48 rem from gamma rays? [*Ans.* 16 rad]

20. A person receives in one day an absorbed dose of 5 rad to the hands in working with an alpha-emitting material. The next day he receives 2 rad whole-body gamma radiation. Assuming that $QF = 20$ for the alpha radiation, calculate the total dose equivalent to this person's hands from the two days' exposure.

21. Based on Table 8-3, show that the maximum permissible cumulative dose equivalent DE to the blood-forming organs of a radiation worker can be expressed as $DE = 5(N - 18)$, where N is the worker's age in years.

22. Which is of more significance genetically: a dose equivalent of 5 rem received by 100,000 atomic energy workers or 0.05 rem received by every member of the world's population (~3 billion)? Explain.

23. A facility is proposed for fabricating radioisotopes for use in cancer research. The site is to be near a city of 100,000 inhabitants. The presence of the facility may raise the background radiation in the city by as much as 5 per cent. The maximum credible accident that could happen in the plant might give an average whole-body dose equivalent of 5 rem to every inhabitant. Discuss the balancing of risk and benefit in deciding whether to permit the building and operation of the facility.

24. Discuss how the balancing of risks and benefits in the last problem affects the cost of treatments involving use of the radioisotopes to be produced.

25. The needs of a growing world population and its increasing per capita power consumption cannot be indefinitely met by the available supply of fresh water and known reserves of conventional fuels (coal and oil). The vast resources of nuclear energy, on the other hand, could be used for a long time to produce heat for distilling salt water into fresh while simultaneously producing electrical power. What risks and benefits are involved in the operation of a nuclear powered desalination plant? Describe the risks to mankind if such plants are not constructed. Suggest possible guidelines for the containment of radioactive materials within proposed plants.

CHAPTER NINE

X-Ray Technology

Never before or since has a strictly scientific discovery become famous so quickly; the nearest rival would be the discovery of nuclear fission in the last days of 1938, but even then the outside world was slower to realise the possibilities.*

G. P. Thomson

9-1 HISTORICAL INTRODUCTION

Roentgen accidentally discovered X-rays on November 8, 1895 while investigating the fluorescence of various materials in gas discharges. Seldom has a scientific discovery had such immediate, widespread, and lasting effect. On the purely scientific level, the discovery paved the way for detailed understanding of atoms (particularly the inner electronic structure), and the use of X-ray diffraction has led to basic new knowledge of the structure of condensed matter. Roentgen lost no time in investigating the properties of the new radiation, the most spectacular of which is its ability to penetrate matter. This property, along with its significance

*G. P. Thomson, *J. J. Thomson and the Cavendish Laboratory in His Day*, Doubleday, New York, 1965. Quoted with permission of the publisher.

Historical Introduction 119

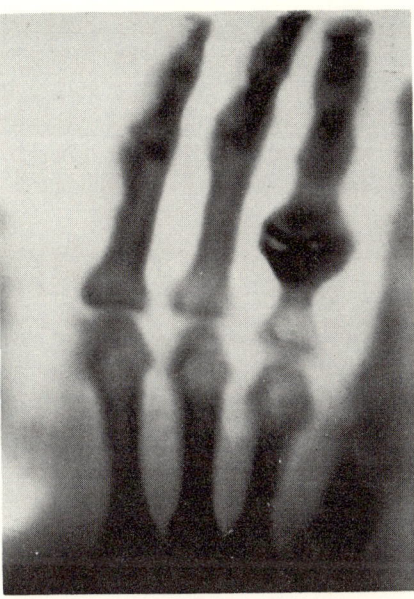

Fig. 9-1 Radiograph of human hand exposed by Roentgen in 1895. (Photo courtesy A. G. Siemens Co., Munich, Germany).

to medicine and industry, was shown dramatically by Roentgen when he succeeded in making a radiograph of a human hand (Fig. 9-1).

Within a short time after Roentgen's discoveries, X-rays were being widely adopted by the medical field as a new and powerful diagnostic tool. Thus, it is not surprising that the X-ray tube itself quickly evolved into a form which is quite similar to those in use today. The authors are privileged to have access to a collection of some of the early tubes assembled by Professor M. L. Pence at the University of Kentucky. Some of these are shown in Fig. 9-2. The tube in Fig. 9-2A, known as a Crooke's tube, is very similar to the one used by Roentgen to study gas discharges when he accidentally discovered X-rays. These tubes contained gas at low pressure which was ionized by means of a high potential applied between the cathode and anode (located in the small side arm of the tube in Fig. 9-2A). The electrons created from the gas ionization process are accelerated to high energy in spite of frequent collisions with the gas, and produce X-rays when they strike the glass at the large end of the pear shaped tube.

Roentgen communicated his discovery of the penetrating properties of X-rays and pointed out their diagnostic potential to the Physico-Medical Society of Wurzburg in December 1895. Within weeks X-rays were being

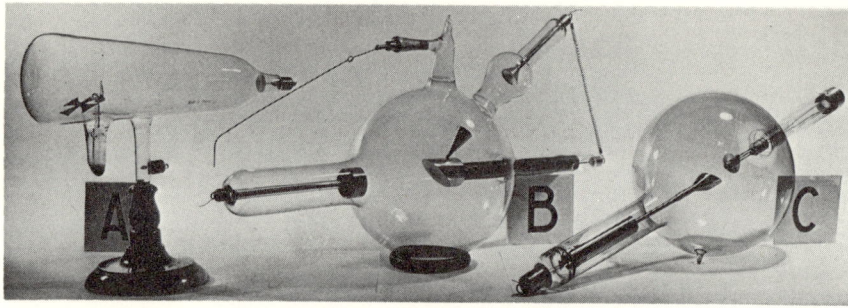

Fig. 9-2 Some of the X-ray tubes used by Professor M. L. Pence in early investigations at the University of Kentucky. Tube A is similar in shape and electrode arrangement to the tube used by Roentgen when X-rays were discovered. Tube B incorporates Professor H. Jackson's development of the concave cathode, which focuses electrons on a smaller area of the anode, set at 45° to increase penetration of the X-rays through the tube itself. Tube C is an evacuated tube and uses a hot filament to produce electrons. This development was due to W. D. Coolidge. This type of tube is widely used today.

used as a medical diagnostic aid. In spite of the fact that the original tubes had several disadvantages, they were widely copied. Experimenters directed their attention to the solution of the many problems encountered with the Roentgen tubes: long exposure times, limited tube life, and lack of image definition caused by the fact that X-rays are emitted from a large glass surface.

In 1896, Professor H. Jackson of King's College, London, developed a fixed focus X-ray tube, of which Fig. 9-2B is representative. In this tube the cathode is concave and electrons are focussed on the anode, which was made of tungsten and served as the target. Jackson set the target at 45° so that X-rays could more easily escape from the tube. He found that the intensity of X-rays was greatly increased because of better focussing, an improved target material, and the improved tube geometry. Another main advantage of this tube is that X-rays originate over a smaller area and thus picture definition is much improved.

The next major advance in the design of X-ray tubes was made by W. D. Coolidge in 1913. A tube of the Coolidge design is shown in Professor Pence's collection, Fig. 9-2C. Appreciation of the contribution made by Coolidge requires an understanding of some of the properties of electrical discharges in gases. Early workers referred to low pressures in an enclosure as a "soft" vacuum and very low pressures as a "hard" vacuum. In the "soft" vacuum region a smaller potential difference is required to maintain the gas discharge than is the case in the "hard" vacuum. Consequently, electrons have greater energy on reaching the target in the "hard" vacuum, and the X-rays emitted are of shorter wavelength and are more penetrating in matter. Even to this day X-rays are classified as hard or soft depending on their wavelength. It is not difficult to imagine that the "quality" of X-radiation emitted from a gas tube depends on the age and previous history of the tube. Furthermore, gas discharges are notoriously variable and hard to control. Coolidge built a tube which was highly evacuated and used a hot filament to produce electrons at the cathode. In the Coolidge tube the number of X-rays produced and their energy can be independently varied. The filament temperature and the target-to-cathode potential difference determine the number of X-rays emitted per second, but the potential difference and the target material determine the energy spectrum of the X-rays. For a given target and potential-difference, the X-ray intensity is controlled by merely changing the magnitude of the filament-heating current.

As an illustration of the rapid spread of the application of Roentgen's discoveries, we note that Professor Pence's records show that many X-ray photographs were taken for Lexington, Kentucky physicians as early as 1903. Figure 9-3 is an interesting example of one of the arrange-

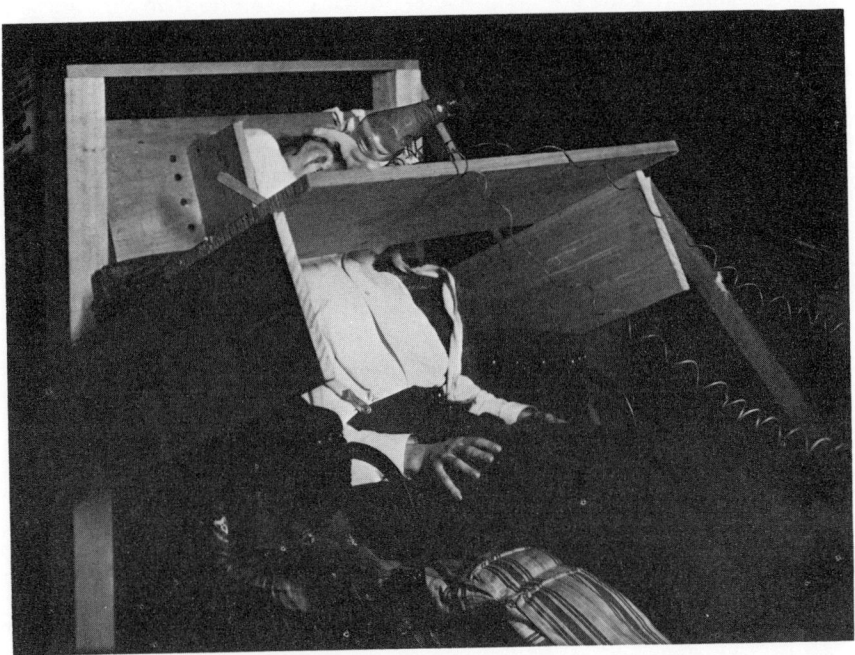

Fig. 9-3 A radiographic arrangement employed by Professor Pence (for Lexington physicians) in the early 1900's.

122 Historical Introduction

ments employed at that time. The tube in use here is remarkably similar to the tube in use by Roentgen when the X-ray was discovered. Professor Pence also applied the new radiographic art to another matter of his locale (Fig. 9-4).

9-2 X-RAY INTENSITY

The penetrating power of X-rays produced by a tube can be changed by adjusting the high voltage, or potential difference, between cathode and anode. The intensity is controlled by the low voltage, which regulates the temperature of the cathode and hence the current of electrons it releases to the anode (target current). The spectrum of X-rays that emerge from a machine depends on a number of factors, such as the construction of the anode, filters, shielding, and tube housing; the high-voltage wave form; and the direction of observation. The amount of attenuation of typical broad beams is illustrated by the measurements summarized in Fig. 9-5. The ordinate gives the exposure rate for 1 ma of target current at a distance of 1 m from a typical target as a function of the thickness of lead through which the beam has passed. For this case, the high voltage was obtained by rectifying 60 cycles ac, producing pulses of 1/120 sec dura-

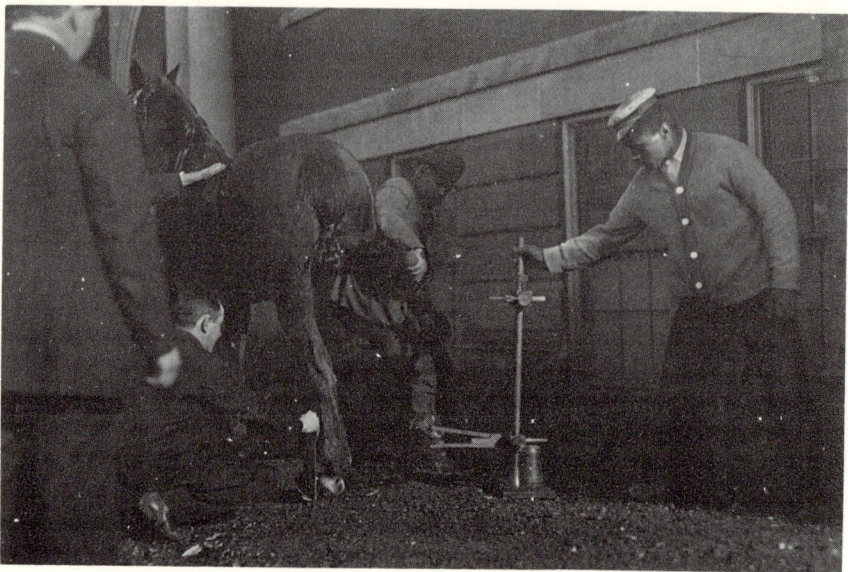

Fig. 9-4 Radiographic arrangement employed by Professor Pence, particularly applicable to his locale.

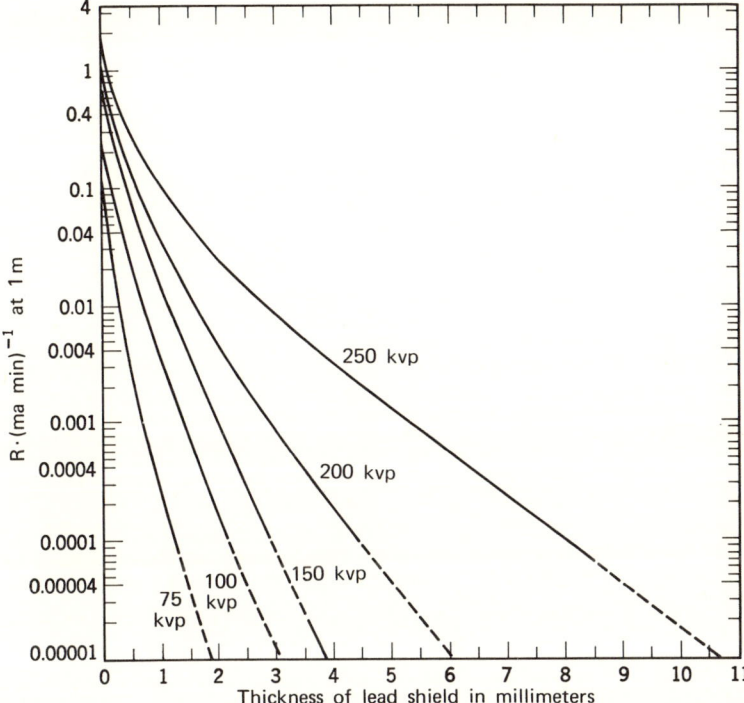

Fig. 9-5 Attenuation of X-rays in lead. (From *X-Ray Protection Design*, National Bureau of Standards Handbook 50, U.S. Department of Commerce, Washington, D.C., 1952).

tion. Curves for different peak voltages in kilovolts (kVp) are shown. In each case the X-ray beams emerged from an anode at 90° with respect to the electron beam and were filtered with an aluminum foil, 0.5 mm thick with the two lowest kVp values, and 3 mm thick with the highest three.

As discussed more fully in the next section, curves such as those in Fig. 9-5 can be used to determine shielding requirements for X-ray beams. The exposure rate from a given machine is directly proportional to the target current and, if the beam is narrow, inversely proportional to the distance from the target.

9-3 SHIELDING AGAINST X-RAYS

An X-ray tube emits X-rays from the target in all directions. On the other hand, radiation is generally required in only one direction, usually in a cone of relatively small solid angle. Let us designate the radiation emitted into the cone of interest (i.e., the radiation which will be used for diagnostic or therapeutic purposes) as *useful radiation*. All other radiations will be designated *nuisance radiation*. Thus, nuisance radiation can serve no useful purpose, it can expose the patient in areas where radiation is not required, it can harmfully expose the physician, it can fog photographic film, and it can interfere in other ways with equipment and instrumentation. Obviously, it is highly desirable to shield against all nuisance radiation, and fortunately this can be done in economical ways with common shielding materials.

Figure 9-5 includes curves for various values of kVp, which is the value of the peak potential applied to the X-ray tube. These values are not to be confused with the X-ray energy. In fact, they represent only the maximum possible value of the X-ray energy, and by reference to Chapter 1, we see that characteristic lines and a continuous spectrum of X-rays are emitted.

It is instructive to comment on the curves in Fig. 9-5 in light of the data presented on X-ray penetration in Chapter 4. We note that in Fig. 9-5 the curves, plotted as ln dose against thickness, are not straight lines as they would be if the exponential absorption law held true exactly. The main reason for the curvature of the lines is that the X-ray beam gets "harder" as it passes through the lead shield. In other words, when a spectrum of X-rays passes through matter, the lower energy portion of the spectrum will be removed more rapidly than the higher energies. This statement holds true when the absorption coefficients are smaller at the higher energies; e.g., Fig. 4-10 shows that this is the case for X-rays in lead unless the photon energy exceeds 2 MeV. After considerable beam "hardening," the curves become approximately straight lines and the exponential absorption law can be used to estimate the "effective energy" of the beam. For example, in Fig. 9-5 we see that the 250 kVp curve is straight between 4.5 and 8.0 mm of lead. Thus

$$\frac{I}{I_0} = \frac{0.0001}{0.002} = e^{-\mu(0.8-0.45)}$$

from which $0.35\,\mu = 3$, and $\mu = 8.6\text{ cm}^{-1}$. This gives $\mu/\rho = 8.6/11.35 = 0.76\text{ cm}^2\text{g}^{-1}$, and an effective energy of about 200 keV, as read from Fig. 4-10. Therefore, the beam contains only the most energetic X-rays after penetration through a few millimeters of lead.

The practical value of the data shown in Fig. 9-5 is easily illustrated. Suppose that, in a medical installation, an X-ray tube is to be operated at 200 kVp and at a current of 1 ma. We see that the dose rate at 1 m from the target exceeds 1 R/min. If it is recommended that the dose (to the operator of the equipment) not exceed 0.3 R in any one week, how much shielding is required? For a 50 hr workweek 0.3 R/week corresponds to 10^{-4} R/min, and 4.5 mm of lead will reduce the dose rate to the required level. Actually, the operator may be closer to the target during part of his work, and it would be advisable to use a slightly thicker lead shield. Maximum economy is affected by installing the lead shield near or on the X-ray tube itself, leaving an aperature for the useful beam.

9-4 X-RAY DOSIMETRY

In Chapter 7 we discussed the principles of measurement of absorbed dose in a medium. The following techniques, which are based on these principles, are frequently used to monitor exposure from X-rays. We shall discuss ionization chambers and film dosimeters here, since other techniques, such as scintillation and glass dosimeters, are used to a lesser extent with X-rays.

Ionization Chambers

The free-air ionization chamber, represented in Fig. 9-6, is the standard instrument for measuring X-ray exposure rates (in R units) in the photon energy range from about 50 to 200 keV. The principle of operation of this chamber can be understood by considering a parallel X-ray beam. An electric field of from 50 to 100 V/cm strength is established by putting a potential difference between parallel plates, P_1 and P_2. The latter plate, being at a positive potential, collects electrons produced as a result of the passage of the X-ray beam through the shaded region, V (the volume traversed by the beam directly above P_2) of the chamber. The plate P_2 is surrounded by a grounded guard ring, G. The separation of P_1 and P_2 must be sufficient to allow all of the electrons released in V to complete their paths in air. Furthermore, electronic equilibrium must exist in V, that is, the number of electrons at various energies leaving the space

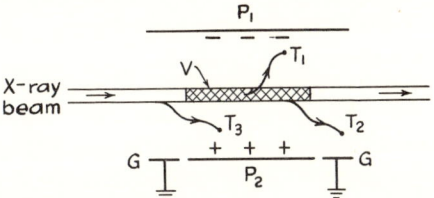

Fig. 9-6 Free-air ionization chamber for measuring X-ray exposure rate. Two conditions have to be met for operation. (1) Electron tracks, such as T_1 and T_2, originating in V must end in air. (2) A track from V such as T_2, which leaves the space directly above P_2, must be compensated by a similar track T_3 that enters this space (electronic equilibrium).

directly above P_2 must be compensated by electrons entering this space. Under these conditions, the number of electrons collected by P_2 is equal to the number of ion pairs resulting from the absorption of X-rays in the volume, V. With a steady X-ray beam the current at P_2 is, by definition, proportional to the exposure rate in V. Knowledge of V and the air density then provides an absolute measurement of the exposure rate in roentgens per unit time. An accuracy of about ±1% can be obtained with the free-air chamber. Its range of usefulness is limited at high photon energies by the requirement that the plate separation be greater than the range of secondary electrons, and at low energies by the lack of electronic equilibrium caused by attenuation of the X-ray beam as it passes through the chamber.

Other ionization chambers, based on the Bragg-Gray principle, are often used to monitor X-ray exposures. Different wall thicknesses are used for different qualities of radiation. Tissue-equivalent wall materials and gases can be employed for measuring absorbed dose.

Pencil ionization chambers that can be carried in a pocket are convenient personnel monitors. The chamber contains an electroscope which, before exposure, is charged by insertion into a suitable instrument. One leaf of the electroscope is a quartz fiber that can be seen against a scale through a simple microscope built in the pencil. When the chamber is exposed to radiation, the electroscope discharges as ions are collected. The movement of the fiber across the scale can be read to determine the exposure. Calibration of the scale in roentgens or milliroentgens is accomplished by using standardized sources.

Films

A small piece of photographic film, enclosed in a light-proof wrapping, provides a convenient X-ray dosimeter that can be worn on the clothing. Films developed every week or month provide a permanent record of exposure. Calibration of the film is done by exposing it to known amounts of radiation and measuring the resulting amounts of blackening when developed. Since the degree of blackening depends on photon energy, film holders often contain several filters of different materials and thicknesses covering different portions of the film's surface. The relative amounts of blackening behind each filter can be used to infer the approximate photon energy spectrum and absorbed dose to the film wearer.

Films must be processed before radiation exposure can be evaluated from them. Pocket ionization chambers, on the other hand, offer the advantage of immediate reading. The accuracy of film dosimetry is typically ±15%, which is considered adequate for routine monitoring of X-ray exposures.

9-5 RADIOGRAPHIC TECHNIQUES

The objective of a radiological exposure is to gain information on the shape, size, location, or condition of organs or objects not normally seen with visible light. As with ordinary photography, the quality of the image produced depends on a number of factors. The most serious of these factors are scattered photons and large X-ray source area which tend to defocus the image. Another problem is connected with obtaining the proper contrast between dense and light materials. Optimization of X-ray energy, use of shields to eliminate X-rays from all directions except in the useful cone, and the use of X-ray tubes which have a sharp focus can substantially reduce the severity of these problems. In some of the newly designed X-ray tubes, rotating targets are used so that the focal point, though fixed in space, describes a circle on the target. In this way very small focal points are produced without overheating the target.

One essential objective of radiography is to produce an image containing the necessary information under conditions where the exposure dose received by the individual is small. Several systems have been developed to accomplish this objective. The most direct method of radiography involves recording the transmitted X-ray intensity with radiographic film. However, the direct use of film requires generally unacceptable exposure levels. For this reason, luminescent screens are used in which the energy absorbed by the screen from the X-ray beam is converted to visible or near visible light. Because each photon in the near visible region carries much less energy than the X-ray photons, many more near visible photons are emitted. This gain in number of photons can be as large as 1000, hence luminescent screens can be used to materially reduce the X-ray exposure. After conversion to visible or near visible photons (with calcium tungstate screens the wavelength is about 4300 Å and with silver activated zinc sulfide the wavelength is about 7600 Å), the image may be viewed directly, as in fluoroscopy. Alternatively, the converted light may be recorded with film, as in radiography, or with photoelectric scanning and electronic devices (image intensifiers). By far the most usual diagnostic radiograph is taken with radiographic film following a fluorescent screen. With a calcium tungstate screen and the more sensitive film, satisfactory images are obtained when the exposure dose measured with the film lies in the region of 1 or 2 milliroentgens (mR). Without the fluorescent screen the same image would require 50–100 mR.[1]

---

[1]Much more information on the techniques of radiography may be found in a recent book *The Physical Aspects of Diagnostic Radiology* by Michel M. Ter-Pogossian, Hoeber Medical Division, Harper and Row (1967).

9-6 CONTROL OF X-RAY EXPOSURE

As stated in the Foreword to this book, medical and dental X-rays constitute about 45% of the total exposure of persons in the United States to ionizing radiation. Therefore, X-rays are the greatest single source of radiation dose to the population at large. Concern for needless exposure and its possible effects on human genetics and on the incidence of certain diseases has grown with recognition of the problem.

Ideally, a medical or dental X-ray machine should be operated so as to achieve an optimum balance between the medical benefit obtained and the exposure of the patient and other persons in the vicinity. To this end, an X-ray installation should, (1) contain proper shielding, (2) have adequate dosimetry, (3) make advantageous use of various radiographic techniques, and (4) be operated with administrative procedures that minimize needless exposure.

These considerations are illustrated by a suggested plan, shown in Fig. 9-7, for a 250 kVp, 15 ma deep therapy installation.[2] Numbers in the figure represent the positions of personnel; and letters show protective barriers. For the purpose of design, it is assumed that, during a 48 hr workweek, the tube will be operated at its full rating and that a maximum exposure of 0.3 R will be allowed to personnel. If there are no mechanical restrictions on the direction in which the X-ray beam can be pointed, then all of the walls, the ceiling, and the floor must contain adequate shielding. By using the 250 kVp curve in Fig. 9-5, we can compute the oblique lead-barrier thicknesses that result in an exposure rate of 0.3 R/week at each numbered position in Fig. 9-7 when the beam is pointed there. The results are shown in Column 4 of Table 9-1. In Column 5, a work factor, depending on the fraction of the time that the beam is pointed in a given direction or that personnel occupy a position, is assigned to each location. Positions 1, 8, and 9, for example, in the hallway were assumed to have a limited occupancy of not more than 6 hr per week. It was also assumed that the beam would not be pointed toward the ceiling for more than one-eighth of the time. Use of this work factor at these four locations reduces the needed oblique barrier thicknesses by three half-value layers [$1/8 = (1/2)^3$] to the thicknesses shown in Column 6. Finally, by considering the angle that the barriers make with the X-ray beam, the thicknesses of the lead shields shown in Column 8 can be found.

Safe and efficient operation of an X-ray facility can best be assured by having a radiation survey made by a qualified person. An expert can give advice on matters such as getting the best pictures, recommended per-

---

[2]This plan is taken from H. O. Wyckoff and L. S. Taylor, *X-Ray Protection Design*, National Bureau of Standards Handbook 50, Washington, D.C. (1952).

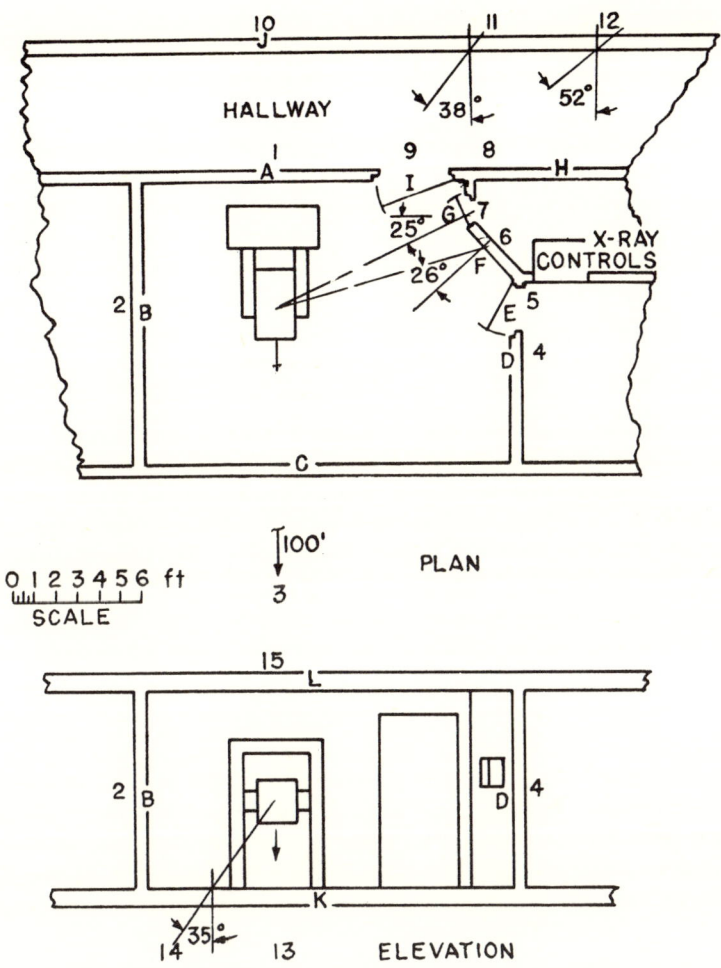

Fig. 9-7 Plan and elevation of an x-ray therapy installation.

sonnel dosimetry, proper operating procedures, and the use of safety interlocks, warning signs, and mechanical restrictions.

PROBLEMS

1. Explain the physical principles involved in making an X-ray picture of a human hand.
2. Why is it important to use a small focal spot in X-ray radiography?

TABLE 9-1 Barrier Protection for Therapy Installation Shown in Figure 9-7

Personnel Position	Barrier	Target Distance (ft)	Lead Oblique Thickness (0.3 R/wk) (mm)	Work Factor	Lead Oblique Thickness (mm)	Angle (Degrees)	Lead Barrier Thickness (mm)
1	2	3	4	5	6	7	8
1	A	6	9.7	1/8	7.3	0	7.3
2	B	7	9.3	1	9.3	0	9.3
3	C	100	3.2	1	3.2	0	3.2
4	D	12	8.2	1	8.2	10	8.1
5	E	12	8.2	1	8.2	0	8.2
6	F	11	8.4	1	8.4	26	7.6
7	G	10	8.6	1	8.6	25	7.7
8	H	12	8.2	1/8	5.8	55	3.3
9	I	10	8.6	1/8	6.2	43	4.5
10	A+J	13	8.0	1	8.0	0	8.0
11	I+J	17	7.3	1	7.3	37	5.8
12	H+J	20	6.9	1	6.9	51	4.3
13	K	8	9.0	1	9.0	0	9.0
14	K	9	8.7	1	8.7	37	7.0
15	L	7	9.3	1/8	6.9	0	6.9

3. List at least three major steps taken in the development of modern X-ray tubes.

4. List at least four advantages of the Coolidge X-ray tube over gas filled tubes.

5. In light of modern knowledge, criticize any unsafe practices evident in making the X-ray picture in Fig. 9-4. Which of the hazards were known before 1895?

6. The exposure rate at a distance of 3 ft from the target of an operating X-ray machine is 2.5 R/hr. What will be the exposure rate if the target current is increased by a factor of 10? At what distance will the exposure rate then be 2.5 R/hr? [*Ans.* 25 R/hr, 9.5 ft]

7. Why do the curves in Fig. 9-5 approach straight lines with increasing barrier thickness? Why do the slopes decrease?

8. Assume that for 250 kVp X-rays the LD-50 for mice is 400 R. How much time would be required to deliver this dose to a mouse placed at 10 cm from an X-ray tube operating at 1 ma with no barrier? Use the data in Fig. 9-5. [*Ans.* 1.33 min]

9. How much time would be required (Problem 8) at a distance of 1 m when a target current of 3 ma is used?

10. Use the data in Fig. 9-5 to calculate the effective energy of 200 kVp X-rays when the barrier thickness exceeds 3 mm of lead. [*Ans.* 150 keV]

11. The volume, V in a free air chamber (Fig. 9-6) is 10 cm³ and contains air at 1 atm pressure and at temperature of 300°K. When a current equal to 10^{-15} amp is observed, what is the exposure rate in R/hr? [*Ans.* 0.92 R/hr]

12. If a man is permitted an exposure of 0.1 R per 40-hr workweek, what fraction of his working time could he spend in the radiation field of Problem 11?
[*Ans.* 0.0027]

13. An X-ray technician was wearing a pencil X-ray dosimeter, which was designed according to the Bragg-Gray principle. The chamber contained 0.05 g of air and had a capacitance of 10^{-11} farad. If it was initially charged at 100 V and found to be discharged by X-rays to 50 V, would you prescribe bedrest?

14. What modifications, if any, would you recommend in the shielding given in Table 9-1 if the hallway in Fig. 9-7 were wide enough to accomodate a desk and chair?

CHAPTER TEN

Applications of Radiation and Radioisotopes

If we try to look into the future and we take the optimistic view that mankind may succeed in organizing itself so as to eliminate the fear and danger of atomic weapons, we might speculate as to what may be the development of atomic energy as a constructive force.

Enrico Fermi*

10-1 INTRODUCTION

Prior to World War II, the principal uses of ionizing radiation outside the laboratory were in medicine, radiography, and the luminous dial industry. With the post-war development of atomic energy, radiation has come into widespread use in industry, medicine, and research. Its present applications are so numerous and varied that no complete listing of them can be given here. The following sections describe a selection of uses indicating

*Enrico Fermi, *Collected Works*, University of Chicago Press, Chicago (1965). Quoted with permission.

the scope and extent to which the unique characteristics of ionizing radiation are employed in biomedical and industrial fields.

10-2 RADIOISOTOPE TRACERS

In addition to its killing effect on malignant cells (Section 8-4), radiation plays another role in biomedical science. Radioisotopes, taken into the body, enter into metabolic processes in the same way as their stable counterparts. It is thus possible to tag an element by adding a radioactive isotope (tracer) and then, by detecting the radiation it emits, follow its path through the body. Virtually every specialty in medicine has found a use for radioisotopes.

The elimination of radioactive material from the body takes place by means of two processes, the biological turnover of elements and the radioactive decay of the isotopes. As discussed in connection with Eq. 2-10, the radioactive decay of an isotope can be described in terms of its half-life, which we shall denote here by the symbol τ_r. We can define an analogous biological half-life τ_b for the elimination of atoms of the isotope from the body or from an organ by biological means. It can be shown that the decrease in the number of atoms of a radioisotope in the body by means of the combined mechanisms of radioactive decay and biological turnover takes place with an effective half-life given by

$$\tau_{\text{eff}} = \frac{\tau_r \tau_b}{\tau_r + \tau_b} \tag{10-1}$$

This effective half-life is smaller than either the radiological or the biological half-life. The radiological half-life of ^{203}Hg, for example, is $\tau_r = 46$ days. The biological half-life of mercury in the body is $\tau_b = 10$ days. The effective half-life of ^{203}Hg in the body is, therefore, by Eq. 10-1, $\tau_{\text{eff}} = 8.2$ days.

Table 10-1 lists some data for radioisotopes in the body and in several organs. The kinds of radiation shown are the principal ones, including rays emitted by radioactive daughters. Only beta and gamma or characteristic X-radiation from within the body can be detected externally. The presence of alpha emitters in the body (e.g., in the bone) can sometimes be inferred from the radiation of their daughters or else detected in body wastes.

10-3 DIAGNOSIS AND THERAPY WITH RADIATION

Diagnosis of metastasized cancer of the thyroid gland can be made by the method illustrated in Fig. 10-1. A patient drinks a solution containing

Table 10-1 Data for Half-lives of Radioisotopes in Various Organs[a]

Radio-isotope	Principal Radiation (Including That from Daughters)	Organ	τ_r (Days)	τ_b (Days)	τ_{eff} (Days)
$^{3}_{1}H$	β	Whole body	4.5×10^3	12	12
$^{14}_{6}C$	β	Whole body	2.0×10^6	10	10
$^{24}_{11}Na$	β, γ	Whole body	0.63	11	0.60
$^{32}_{15}P$	β	Whole body	14.3	257	13.5
		Liver		18	8
$^{40}_{19}K$	β, γ	Whole body	4.6×10^{11}	58	58
$^{42}_{19}K$	β, γ	Whole body	0.52	58	0.52
		Muscle		58	0.52
$^{45}_{20}Ca$	β	Whole body	164	1.64×10^4	162
		Bone		1.8×10^4	162
$^{59}_{26}Fe$	β, γ	Whole body	45.1	800	42.7
$^{68}_{31}Ga$	β^+, γ	Whole body	0.047	6	0.047
$^{90}_{38}Sr$	β, γ	Bone	10^4	1.8×10^4	6.4×10^3
$^{131}_{53}I$	β, γ	Whole body		138	7.6
		Thyroid		138	7.6
$^{198}_{79}Au$	β, γ	Whole body	2.7	120	2.6
$^{226}_{88}Ra$	α, β, γ	Bone	5.9×10^5	1.64×10^4	1.6×10^4
$^{239}_{94}Pu$	α, β, γ	Bone	8.9×10^6	7.3×10^4	7.2×10^4

[a]Report of the ICRP Committee II on Permissible Dose for Internal Radiation (1959), *Health Physics*, Vol. 3 (1960).

radioiodine, ^{131}I, which, after beta decay, emits gamma rays ranging in energies up to 0.722 MeV. Almost all of the iodine going into the bloodstream becomes concentrated in the thyroid, as is easily verified by detecting the emitted gamma radiation with a Geiger-Mueller (GM) tube or scintillation counter. The presence of a metastasized thyroid cancer shows itself by the presence of radioiodine in the cancerous off-shoots (metastases) extending out of the gland. In addition to its importance in diagnosis, ingested radioiodine is also used therapeutically to destroy cancerous tissue in the thyroid.

Hyperthyroidism can be diagnosed by observing the rate at which tagged iodine goes to the thyroid after being drunk. As a rule, a normal thyroid will take up less than 20% of the ingested iodine during the first hour, while a hyperactive gland may accumulate twice this amount. Radioiodine is also used to treat hyperthyroidism and certain cases of heart disease by destroying thyroid tissue. Because of its shorter half-life (6 hr) and its lower-energy gamma radiation, metastable (see Section 2-2) technetium (^{99}Tc) is often preferred in making thyroid scans and radiocardiograms.

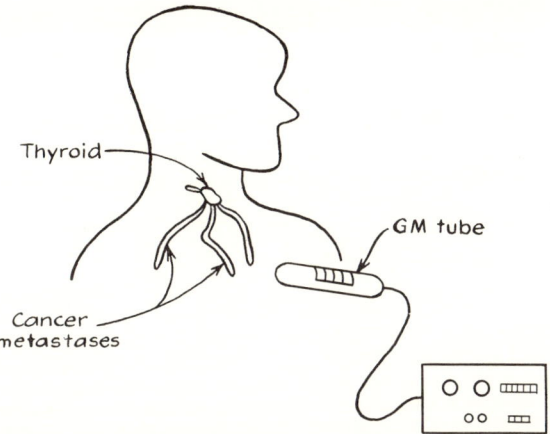

Fig. 10-1 Diagnosis of metastasized cancer of the thyroid by means of radioiodine. The isotope is also employed therapeutically.

Rabbit antibodies to human fibrinogen, injected intravenously, show considerable specificity for localization in tumors containing dark pigment (melanomas) and in certain carcinomas. These antibodies, labeled with radioiodine, are therefore useful in the diagnosis of tumors.

The localization of specific chemicals in particular organs, and in certain abnormal growths has been exploited in the diagnosis and treatment of a large variety of other disorders. Neurosurgeons, for example, in diagnosing and locating brain tumors, have injected radioactive arsenic (^{74}As) and gallium (^{68}Ga), which are rejected by normal brain tissue. An emitted positron from these isotopes annihilates with an electron to create two 0.51 MeV gamma rays, which travel in opposite directions. The site of the tumor can be traced from the observed lines of travel of the gamma ray pairs, as detected by an array of coincidence counters. Other radioisotopes are used to diagnose and treat abnormal growths elsewhere in the body. By selecting isotopes that emit radiation with the right energies and ranges, certain tissues or organs can be selectively destroyed, leaving adjacent tissues with little or no damage. In many cases therapeutic implants of radioactive gold (^{198}Au), yttrium (^{90}Y), phosphorous (^{32}P), palladium (^{103}Pd), and radium (^{226}Ra), are made.

Charged particle accelerators have been employed to produce external radiation for treatment of tumors without surgery. As shown in Fig. 3-4, the stopping power curve for protons, like that for other charged particles, peaks very sharply at the end of the range. When the position of a brain tumor, for example, has been located, then with the proper amount of material placed in the path of the beam, the charged particles come to rest

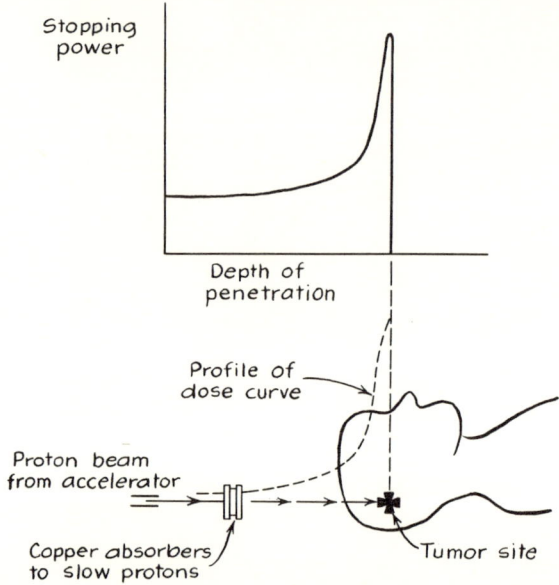

Fig. 10-2 Tumor destruction with charged particles. Particles brought to rest in the tumor deposit their peak dose there. Absorbers of copper or other materials slow the beam down to the desired energy as it enters the head.

in the tumor. As illustrated in Fig. 10-2, the peak local dose is thus absorbed in the tumor and not in the adjacent tissues. The selective destruction of the tumor can also be enhanced by applying the particle beam from several different directions of incidence.

Radioactive ^{60}Co is extensively used externally in place of X-ray machines to produce radiation for cancer treatments. Its energetic (1 to 2 MeV) gamma rays are more advantageous than X-rays, having nearly uniform energy and greater penetration power. A radiocobalt therapy unit is shown in Fig. 10-3.

The isotope ^{60}Co is also used in the diagnosis of certain blood disorders. Vitamin B-12 tagged with ^{60}Co is administered orally and the urine checked for radioactivity. The slower than normal clearance of vitamin B-12 from the digestive tract to the blood in patients with chronic myelogenous leukemià enables diagnosis of this disease to be made with the ^{60}Co labeled vitamin.

Radiophosphorous (^{32}P) and other isotopes are used in the treatment of a number of blood diseases, such as chronic leukemia and *polycythemia vera*. A large portion of phosphorous, drunk in a water solution, goes to the bone marrow where its radiation inhibits blood cell production.

Diagnosis and Therapy with Radiation   137

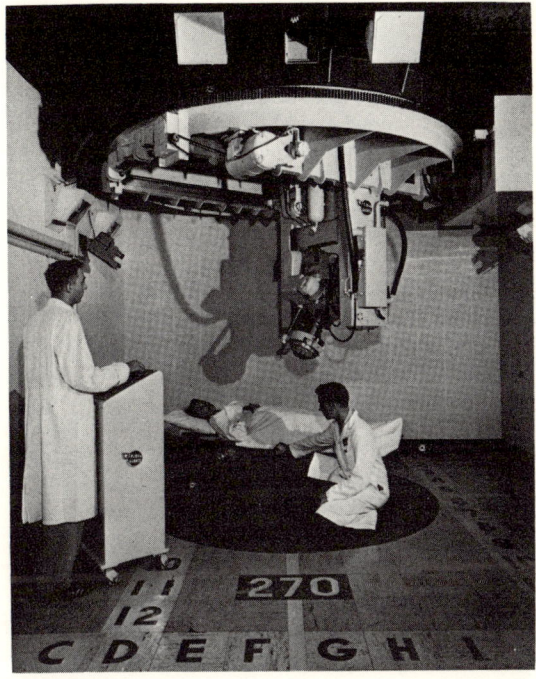

Fig. 10-3 Radiocobalt radiotherapy installation. (Photo courtesy Medical Division, Oak Ridge Associated Universities).

Radioactive strontium (^{90}Sr) is used to treat benign tumors on the white of the eye. The low-energy beta rays (up to 0.55 MeV) do not penetrate the eye and hence do not injure deeper lying tissue. Tumors placed in contact with a ^{90}Sr source can be destroyed without surgery.

The use of colloidal suspensions of radiogold (^{198}Au) to diagnose a number of diseases is illustrated in Fig. 10-4. The scans are made with an instrument in which a radiation detector transmits to a recorder countrates, showing the relative concentrations of radioactivity at successive locations. Overactive bone marrow readily picks up intravenously injected colloidal gold, as seen from the contrast that appears in the cases shown in Fig. 10-4. A similar scanning procedure utilizes ^{131}I, as described by Fig. 10-5, to detect surface lesions.

These examples by no means exhaust the diagnostic and therapeutic uses of radioisotopes. Other important applications include the uses of injected ^{24}Na in radiocardiography and in locating normal and restricted blood flow elsewhere in the body; radioactive mercury to study kidney,

138 Applications of Radiation and Radioisotopes

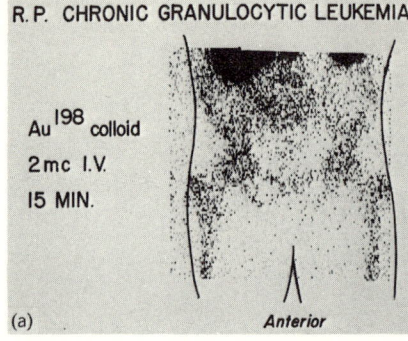
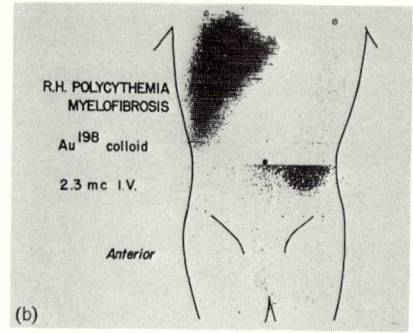
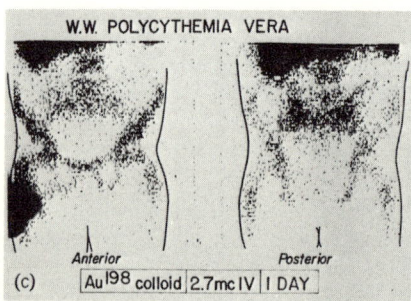
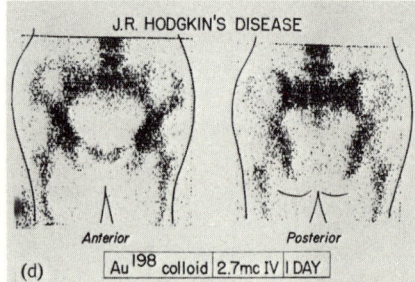

Fig. 10-4 Bone marrow scans (Oak Ridge Associated Universities). Because bone marrow readily picks up colloidal gold after intraveneous injections of radioactive ^{198}Au, scans can be made of the marrow to relate its distribution to a particular disease. In the scanning process, the radiations from the isotope are picked up by sensitive detectors passed at constant speed over the body. The count rate is continuously transformed into electronic impulses that print dots with a frequency proportional to the count rate. (a) Uptake of intravenous colloidal ^{198}Au in the marrow, by a patient with chronic granulocytic leukemia. (b) is a scan of a patient with polycythemia and myelofibrosis. Although an enlarged spleen is shown, there is no evidence of ^{198}Au in the pelvic bone marrow; this finding suggests that the marrow has been replaced by fibrous tissue (myelofibrosis). (c) is a scan of a patient with polycythemia vera, and (d) is a pelvic scan of a patient with Hodgkin's disease. The uptake of the radiogold in the two bottom scans indicates active marrow. (Photo courtesy U.S. Atomic Energy Commission).

liver, and spleen disorders; radioactive selenium in diagnosing cancer of the pancreas; colloidal radiogold for liver scanning; and radioactive fluorine (^{18}F), calcium (^{47}Ca), and strontium (^{85}Sr) for the detection of bone disorders. Other procedures and techniques are being discovered and perfected by continuing medical research.

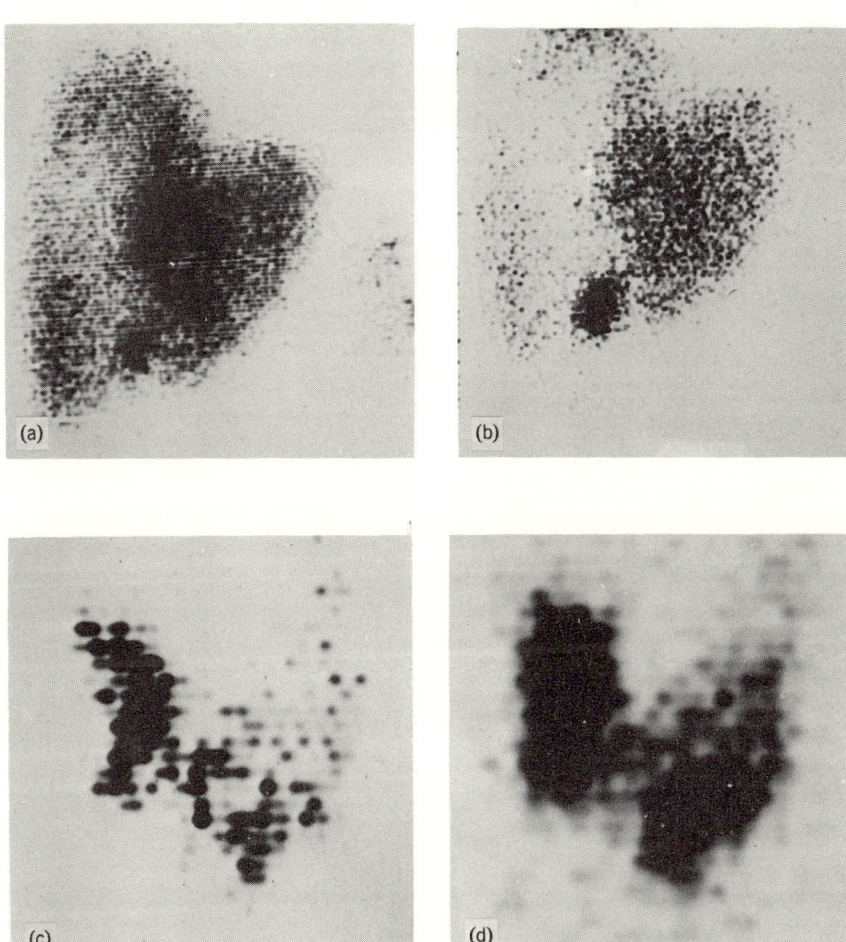

Fig. 10-5 The radioisotope iodine-125 has greatly enhanced the use of radioisotopic scanning procedures because of its longer half-life and very low energy gamma ray, which gives a much clearer resolution, compared with iodine-131 (Argonne Cancer Research Hospital). A liver scan with ^{131}I labeled Rose Bengal dye is shown in a and one with ^{125}I labeled Rose Bengal in b. Comparison shows the increased contrast produced with the latter isotope for the detection of surface lesions. c and d show scans of a multinodular thyroid with 50 μc of ^{125}I and 50 μc ^{131}I. Note the superior definition obtained with ^{125}I even when almost half of the emitted photons are absorbed in the overlying tissue. (Photo courtesy U.S. Atomic Energy Commission).

10-4 MEDICAL STUDIES AND RESEARCH

In addition to their diagnostic and therapeutic uses, radioisotopes are helping to clarify fundamental questions about the functioning of the body and the nature and causes of disease. A few examples will illustrate the vital role that radiation plays in medical research. Experiments with animals are often carried out when studies with humans are not feasible.

Radiocardiographic studies, already mentioned in connection with dianosis of heart disorders, yield important information often not obtainable by other means. The detailed pumping action of the heart, for example, is studied as illustrated in Fig. 10-6. The passage of intravenously injected ^{24}Na through the heart can be traced and recorded with GM counters placed externally at positions of interest. It is thus possible to follow the passage of blood through the heart and lungs. Quantitative studies of the time-rate of cardiac output can be made with intravenously injected human serum labeled with ^{131}I.

Total blood plasma volume determinations are made with iodine-labeled proteins. A known amount of intraveneously injected radioisotope becomes dispersed throughout the circulatory system. The radioactivity in a plasma sample is then compared with that of the injected sample, the dilution of the radioisotope determined, and the total blood plasma volume computed (Problem 9). The total blood volume can then be determined from the hematocrit value, which gives the ratio of plasma and total-blood volumes.

The physiology of iron in the blood is studied by injecting radioactive

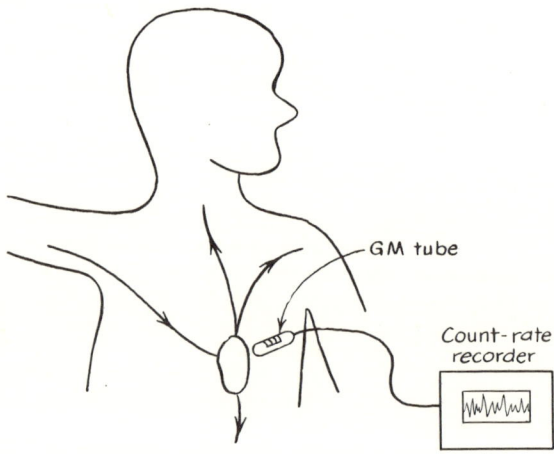

Fig. 10-6 Radiocardiography with ^{24}Na.

^{59}Fe in the form of FeCl$_3$ added to plasma. The total iron in the plasma and its rate of turnover can be determined from analysis of blood samples taken after various time intervals. In addition to yielding other basic physiological information, these studies have shown that the turnover of iron in the plasma is increased in persons with anemia, polycythemia, and leukemia.

From studies with NaCl labeled with ^{24}Na, it has been found that injected sodium reaches some body fluids within a matter of seconds and appears in perspiration within two minutes. The rates of turnover of a number of elements in various body tissues and organs have been simply ascertained with the help of radioisotopes.

Research in intermediary metabolism has progressed rapidly with the help of radioactive tracer techniques. The breakdown and utilization of molecular fragments by the body are being studied by using radioisotope-labeled organic compounds. For example, amino acids tagged with radioactive ^{35}S reveal how they enter metabolic processes and are utilized by various organs. Drugs, hormones, vitamins—virtually any substance—can be labeled. The carbon in organic molecules or molecular fragments can be labeled with ^{14}C, or the hydrogen with tritium, ^{3}H. It is possible to label a specific part of an organic molecule such as a particular acetyl group, and observe how it is handled by the body. In addition to normal metabolic information, these studies show the deficiencies and abnormalities that accompany disease. Such information can often be applied directly in diagnosis and therapy.

10-5 MOLECULAR AND CELLULAR LEVEL RESEARCH

Techniques of organic molecule labeling have led the way to research aimed at understanding how biochemical molecules and other cellular constituents are synthesized and how reproduction of the cellular unit of life from raw materials in the body is carried out. A few examples of such investigations will be given.

There is evidence that the protein, collagen, is dependent for its structural strength, at least in part, on the hydroxylysine it contains. Radioisotope studies have shown that all hydroxylysine in collagen is formed from lysine that adds an atom of oxygen during incorporation into the protein. Injected free hydroxylysine is not synthesized into collagen. To study the specific manner in which collagen synthesis is accomplished by the body, selected carbon and hydrogen atoms in the lysine chain are labeled radioactively. Direct oxidation at a certain carbon atom, for example, then removes tritium in place of a normal hydrogen. Various hypotheses about the specific way in which collagen is formed biochemically can thus be tested.

Investigations of important sulfur containing compounds, their reaction rates, and the discovery of enzymes dependent on them have been made with tracers. These compounds have been found to play highly diversified and specific roles in biochemistry, such as in the synthesis of the methyl group.

The dynamics of mitochondria are studied with radioisotope tracers. The discovery that a number of different proteins in the mitochondrion are replaced at the same time indicates that it is replaced as a unit. The effects of hormone concentration on the activities of enzymes in the mitochondrion can be quantitatively measured with radioisotope tracers.

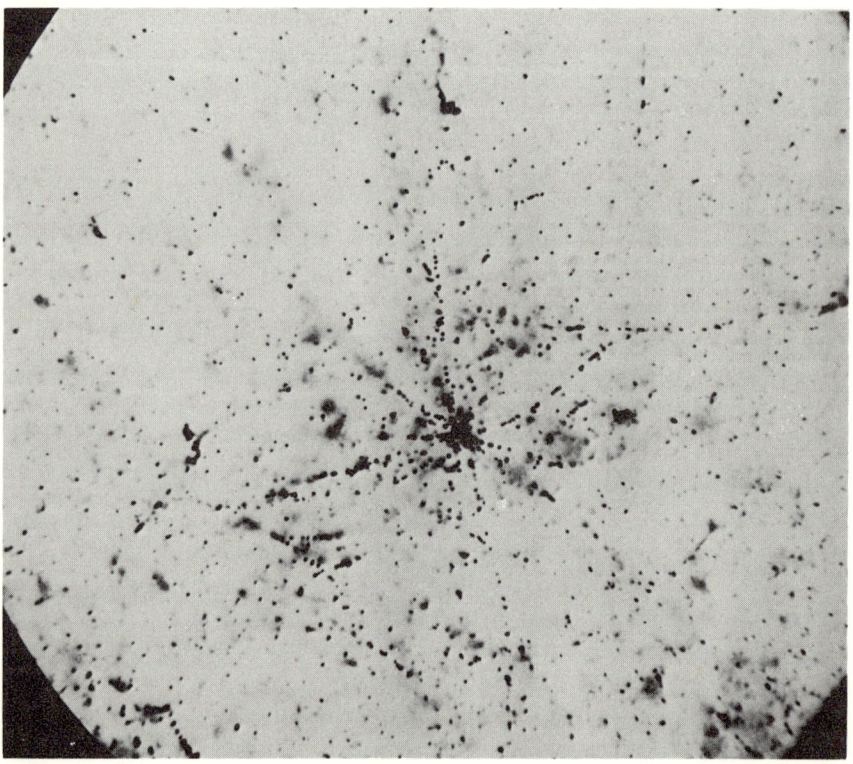

Fig. 10-7 Molecular autoradiograph (Johns Hopkins University). By a careful analysis of the number of rays arising from the "stars" in this autoradiograph it is possible to calculate the number of ^{32}P atoms in the DNA molecule and thus estimate its molecular weight from its known composition. (Photo courtesy U.S. Atomic Energy Commission).

The molecular weight of the DNA molecule has been estimated from autoradiographs, such as the one shown in Fig. 10-7, made from DNA tagged with ^{32}P. By analyzing the number of beta particle tracks on the film, the total number of phosphorous atoms in the DNA molecule was estimated. The molecular weight of the molecule was then determined from the known chemical composition of DNA. It was also found, by means of this technique, that the $T2$ bacteriophage particle is a single molecule with a molecular weight of 1.33×10^8. Since this is one of the largest bacteriophage known, the result suggests that bacteriophage of all types contain single molecules of nucleic acid.

Cell population kinetics has been analyzed with the help of thymidine labeled with ^3H and ^{14}C. By supplying tagged thymidine at various intervals during cell replication and subsequently locating and identifying the radioisotopes autoradiographically, the duration of phases in the mitotic cycle can be determined. Figure 10-8 shows two such radiographs. The photograph on the left shows chromosomes of the root tip of *Tradescantia* treated with labeled thymidine. The two nuclei at the top, with localized grains, were labeled with tritium. The nucleus at the bottom was labeled with ^{14}C. The photograph at the right shows the root tip chromosomes, treated with tritiated thymidine, midway and near the end of DNA synthesis. Following treatment midway during the synthesis, tritium atoms became distributed throughout the arms of the chromosomes, while following treatment near the end of synthesis, they appeared only near the ends of the arms.

10-6 RADIATION GENETICS

The production of gene mutations by ionizing radiation is utilized in two ways. First, radiation is employed to accelerate normal mutation rates, in order to breed selectively organisms with desired characteristics. Second, it is used as an experimental biological "probe," its observed effects giving clues about the genetic control of differentiation and development.

Experiments to develop disease-resistant strains of vegetables, cold-resistant fruits, and genetically stable antigens have demonstrated the importance of radiation as an artificial mutagenic agent. Early ripening peach trees, short-stem grains, high yield peanut and bean varieties, and disease-resistant wheat and oats have been produced. In addition, special strains, such as genetically conditioned plants, susceptible to tumor formation, are bred for biological experimentation.

We can give here only a few examples to indicate the vast scope of radiation studies in basic genetics. The discovery of the dose-rate de-

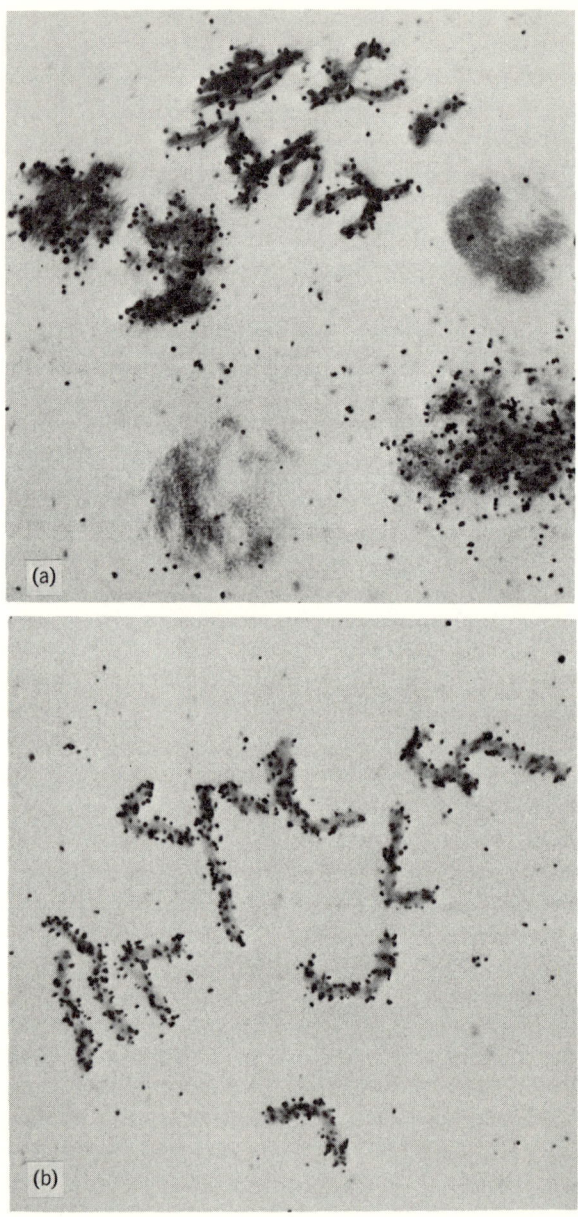

Fig. 10-8 Estimation of cell population kinetics carried out with ^{3}H and ^{14}C labeled thymidine (Brookhaven National Laboratory). See Text. The autoradiograph (a) is magnified 1800 times; and (b) 3600 times. (Photo courtesy U.S. Atomic Energy Commission).

pendence of mutation rates has already been mentioned (Section 8-8). Extensive X-ray studies with *Drosophila melanogaster*, in which over one million individuals were examined, have shown a mutation rate proportional to exposure down to 5 R. Comparative studies of the genetic effects of gamma rays and neutrons help elucidate mechanisms of mutation. To understand how genes function, detailed biochemical studies have been undertaken. Mutations are examined for correlations with changes in DNA and other cellular constituents. It has been found, for example, that some mutations in microorganisms result in amino acid replacements in DNA, which affect substrate affinity. Areas of the molecule concerned with reactions of certain classes of chemicals present in the cell have been identified. In *Escherichia coli B* and other bacterial strains, observation of the synthesis of mutant DNA and the mutation of DNA after synthesis have provided information about genetic factors that control growth and development.

10-7 AGRICULTURE

Tracer studies, similar to those in medical research, are conducted in the fields of animal and plant metabolism and nutrition. As the following examples show, information obtained with the help of radiation has already been of great practical importance.

Testing the efficiency of animal rations, and understanding the value of certain nutrients in diet have been greatly improved through the use of radioisotopes. The digestibility of calcium in various diets, particularly in that of the cow during times of milk production, has been studied. The difference between the total calcium intake and the amount excreted is not a measure of its digestibility, because an animal has a reserve source of calcium (e.g., in the blood and bones). By putting radioactive calcium into the animal's diet, however, it is possible to measure the fraction that is actually digested — typically about 35%. Other nutrients can be similarly studied. Tests with radiophosphorous show that about 20% of the phosphorous in milk, and about 65% of the phosphorous in eggs comes directly from the feed.

Much has been learned about the dietary needs of fetal animals. The path of radiocalcium injected into a cow can be followed by a GM tube outside the cow's body. The calcium demands of unborn calves, at various times during gestation, have been learned in this way.

With some animals the efficiency of the diet in producing lean meat rather than an over-all weight increase is important. Potassium is present in the body, almost exclusively, in the lean meat portions, the amounts in fat, bone, and water being small. The amount of lean meat in an animal, therefore, is approximately proportional to the amount of potassium in its body. Since all potassium contains a small fraction of naturally occurring radioactive potassium, ^{40}K, measuring the count-rate of this isotope from an animal gives a measure of the amount of lean meat in the animal.

The use of hormones to fatten animals and tranquilizers to reduce weight losses during shipping have been proposed as a way of decreasing the cost of meat production. Before such methods may be applied commercially, it must be demonstrated that use of these substances will not be injurious to the consumer who eats the meat. Labeled with tritium, the drugs are detectable in the trace amounts (below the levels of chemical detection) in which they are used, and tests can thus be made to see how much remains in the product to be marketed.

The effectiveness of fertilizers has been improved through investigations with radioisotopes. Scientists have learned from tracer studies with nitrogen, phosphorous, and potassium that many chemicals are taken up directly through the foliage of plants. Nutrient foliage sprays are now used commercially for a number of crops, thus avoiding inherent wastes due to leaching of fertilizers applied to the soil. Other facets of plant metabolism have been discovered with tracers.

Radioisotopes have aided in learning about the chemical and biological control of plant diseases and weeds. They are also used to trace the habits and food chains of insects, some of which interfere with agricultural production, and some of which act as friendly predators of harmful bugs. As with plant diseases, biological control of pests is potentially more economical than artificial control. The screwworm fly has been brought under control in areas of the southeastern United States by releasing to the environment large numbers of flies that were sterilized by radiation. The resulting sterile matings with members of the outnumbered normal male population sharply reduced the size of the next generations. In less than two years the screwworm fly was eliminated in several states where this program was carried out.

Food can be preserved with radiation. Doses of several million rad sterilize food completely. Lower doses keep potatoes and onions from sprouting when kept many months in storage at room temperature.

10-8 ENVIRONMENTAL STUDIES

Ecological, atmospheric, and oceanic studies utilize radiation. Included are researches on biological uptake, mixing, and concentration of chemicals; soil-plant-water relationships; radioactive fallout in soil, food, and man; and atmospheric and oceanographic dynamics. Tracers are used to follow substances through ecological systems. As an added environmental factor, radioisotopes help determine both the effects of widespread radioactive contamination and the countermeasures that are effective. In addition to their importance in basic research, such studies are used for assessing standards of controlled radioactive-effluent release from installations using radioactive materials, for developing safe procedures of large-scale radioactive waste disposal, and for civil defense planning.

10-9 INDUSTRIAL USES

Radiation is employed in both industrial production and research. In construction, for example, gamma- or X-ray photographs (radiographs) of welded seams are often made. In such work the object to be tested is placed between the radiation source and photographic film, which, after exposure, is developed. Like an X-ray photograph of the body, the film shows light and dark areas, depending on the amount of matter traversed by the radiation. Figure 10-9 shows an X-ray radiograph of a defective stainless steel valve casting, which appears normal to ordinary sight and touch. Such radiographs play an important role in the construction of safe buildings and vehicles.

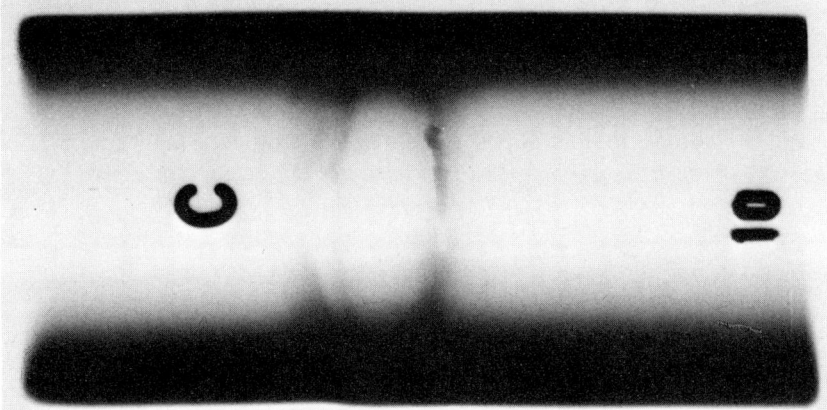

Fig. 10-9 X-ray radiograph (positive) of two pieces of pipe welded together. This radiograph was made by placing film under pipe and irradiating obliquely from above. The light areas show where the most radiation penetrates. A dark region can be seen where the weld protrudes into the interior of the pipe. (Photo courtesy Oak Ridge National Laboratory).

Fine quality control in the manufacture of sheet metal and paper products is possible with radiation thickness gauges. As illustrated in Fig. 10-10, the count rate registered in a GM counter from a radiation source depends on the thickness of sheet metal passing between the two. The signal from the counter can be used to control automatically the pressure of the rollers in order to maintain precisely the desired thickness of the product.

The location of a leak in underground pipes can be found with the help of radiation, without extensive digging. A radioisotope (e.g., radioiodine), introduced into one end of the piping, seeps through the leak and collects in the surrounding earth. The position of the leak can then be determined with a survey instrument above ground.

Neutron sources and gamma-ray detectors, lowered into holes drilled in the ground, are used in exploration for oil. Underground layers of earth, rich in petroleum, are characterized by their relatively high content of hydrogen and low amount of background radiation. Neutrons produced by the underground source are captured by surrounding hydrogen nuclei, which then emit gamma rays, the reaction being $_1^1H + _0^1n \rightarrow {}_1^2H + {}_0^0\gamma$. The presence of oil is evidenced by detection of the gamma ray.

Research with radiation has improved a number of commercial products, not only in quality, but also in lowering cost to the consumer. Tests of tire wear, for example, which formerly required many days of highway driving, can now be made in a matter of a few hours by using tires with radiophosphorous added to the rubber. Before a test is made, the count rate at the surface of the tire is measured. The tire is then run on a machine, simulating contact with a road, and the amount of wear is determined from the subsequent decrease in the surface count rate.

The efficiency of soaps, detergents, and washing machines has been tested by adding radioisotopes to the dirt in clothes. Small amounts of

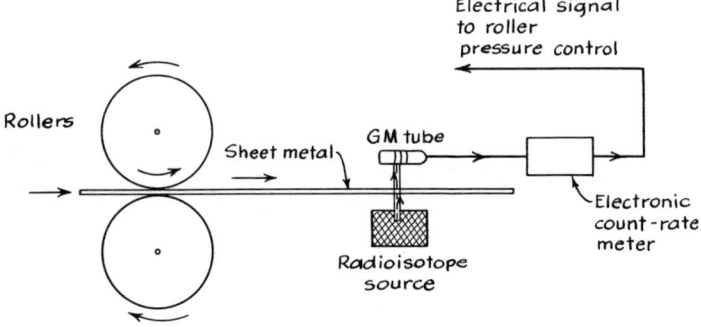

Fig. 10-10 Radiation thickness gauge.

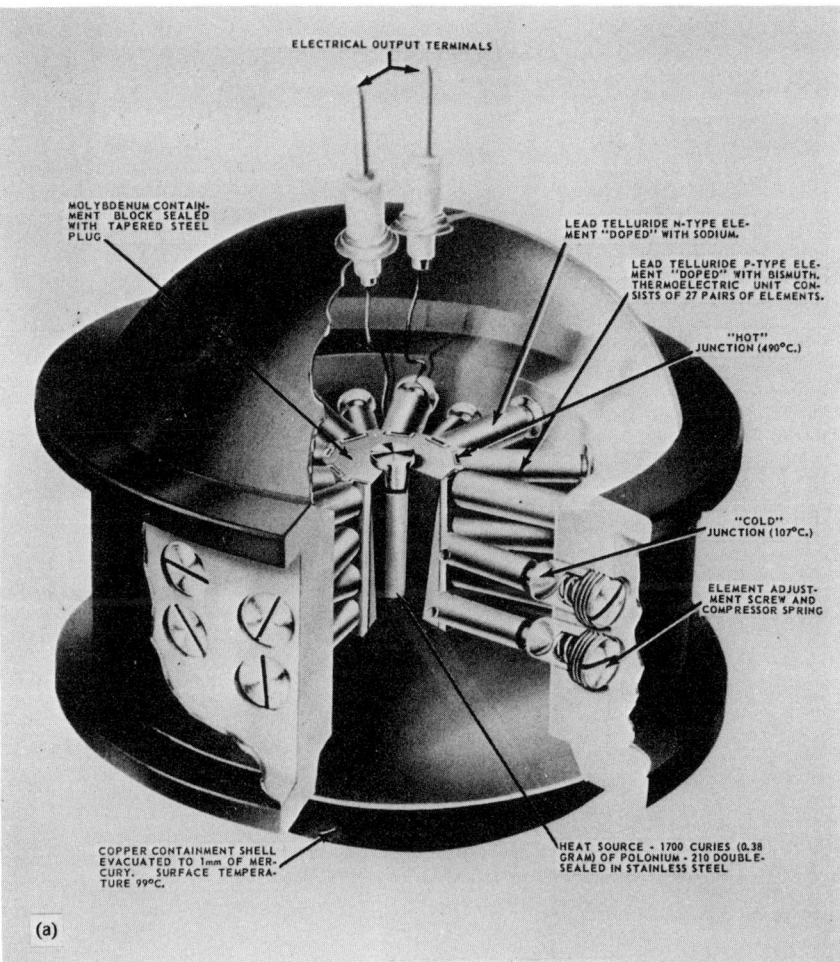

Fig. 10-11(a) and (b) SNAP III thermoelectric generator (The Martin Company). SNAP = System for Nuclear Auxiliary Power. (Photo courtesy U.S. Atomic Energy Commission).

radioactive material remaining on the washed clothes can be easily detected with instruments.

Sensitive tests of engine lubrication and wear are made with radiation. A piston to be tested is put into a reactor, where a fraction of its atoms becomes radioactive through the absorption of a neutron. The stable iron isotope, $^{58}_{26}$Fe, (abundance 0.31%) is made radioactive by the reaction $^{58}_{26}$Fe + $^{1}_{0}n \rightarrow {}^{59}_{26}$Fe. The radioisotope ^{59}Fe decays by emission of a beta

150 Applications of Radiation and Radioisotopes

particle and subsequent gamma rays with a half-life of 45 days. When used in an engine test, the surface of the piston wears down, the worn-off iron collecting in the motor oil. As little as 10^{-5} ounces of iron in oil can be detected through the presence of radioactive ^{59}Fe.

10-10 NUCLEAR POWER

Nuclear reactions release much larger amounts of energy than chemical reactions. The latter involve interactions between outer-shell atomic electrons, which have energies of tens of electron volts or less per atom. The release of millions of electron volts per atom in radioactive decay is common, and nuclear fission releases about 200 MeV per atom. The energy reserves in nuclear fuel deposits far exceed those of fossil fuels. At the present rate of consumption, the world's supply of economical coal and oil could be exhausted within the next century. Nuclear reactors today produce commercial electrical power from controlled fission, and the possible control of nuclear fusion on a large scale is under intense study.

Nuclear power released in radioisotope decay is utilized on a small scale to do a number of jobs not feasible with chemical power. Figure 10-11 shows a thermoelectric generator powered by 1600 curies of the polonium isotope $^{210}_{84}$Po, which emits a 5.3 MeV alpha particle with a half-life of 138 days. The alpha particle travels only a few millimeters, and the kinetic energy it loses by collision with other atoms is converted into heat. One end of a thermocouple is placed in the warm radioactive sample, and the other is placed outside, where it is cooler. The temperature difference causes a current to flow through the thermocouple. The power released by radioactive decay (about 500 watts) is thus converted continuously into electrical energy, the efficiency being about 10%. Radioisotope power generators, which have no moving parts and need no maintenance, have been used as power supplies in earth satellites, navigational beacons, and remote weather stations. Longer-lived isotopes, such as those of strontium and plutonium, have also been used.

PROBLEMS

1. Prove that the effective half-life given by Eq. 10-1 is always smaller than both the radiological and the biological half-lives.
2. A stable isotope can be described by letting τ_r become infinitely large in Eq. 10-1. Show that when $\tau_r \to \infty$ the effective half-life approaches the biological half-life: $\tau_{\text{eff}} \to \tau_b$.
3. Calculate the value of τ_{eff} from the values of τ_r and τ_b for ^{90}Sr in bone given in Table 10-1. [Ans. 6400 yr]
4. The radioisotope ^{24}Na emits a gamma ray of energy 0.45 MeV. Describe a simple experiment to determine the time required for sodium, drunk in the form of salt water, to reach various parts of the body.
5. What dose rate results from the implant of 1 mc of radiopalladium, ^{103}Pd, (half-life 17 days) per g of tissue? This isotope decays by electron capture, emitting, per decaying atom, a total of 0.57 MeV in X-radiation, which can be assumed to be absorbed locally in the tissue. [Ans. 0.34 rad/sec]
6. The use of ^{60}Co is often preferred to X-rays in external radiotherapy because the ^{60}Co gamma rays are more penetrating and are more nearly monoenergetic. Why are these two properties advantageous in patient treatment?

7. Some forms of technetium in the body behave similarly to iodine. The metastable isotope ^{99}Tc has a (radiological) half-life of only 6 hr and emits no particulate radiation. Based on its physical properties, what advantages does this isotope offer over ^{131}I in making thyroid scans?
8. Figure 10-5 shows a comparison of liver and thyroid scans made with the conventional ^{131}I and with ^{125}I. The same scanning instrument was used throughout. The "newer" isotope ^{125}I has a longer half-life and a lower energy gamma ray than ^{131}I. Why do these properties of ^{125}I result in the better scanning resolution?
9. In a blood plasma volume determination, 3 cm³ of iodinated protein containing 100 μc/cm³ of ^{131}I is injected into a patient. Later a sample of blood plasma from the patient shows a radioiodine content of 0.0277 μc/cm³. Estimate the total volume of blood plasma in the patient. [*Ans.* 1.1×10^4 cm³]
10. Discuss how tritiated (labeled with ^{3}H) thymidine could be used in nucleic acid metabolism studies.
11. Design an experiment to test whether fertilizer mixed throughout the soil or just below the roots of a plant is more beneficial.
12. By using radioactive tracers, how would it be possible to determine whether bees cross pollinate two species of plants separated by a distance of ¼ mile?
13. Make a list of organs or systems of the body and the radioisotopes that are used in their study, or in the diagnosis and treatment of disorders in each.
14. How do benefits derived from radiographic analyses of welded joints with ^{60}Co in a ship justify the exposure of individuals during the fabrication and use of the source?
15. Using the data given about the SNAP III generator at the end of Section 10-10, verify that the power produced by the radioactive decay of 1600 curies of $^{210}_{84}$Po is about 500 watts.

APPENDIX A

Historical Outline

c. 1810 Dalton's atomic theory
   1839 Daguerre discovers photography
   1859 Bunsen and Kirchhoff orginate spectroscopy
   1865 Mendel finds basic laws of genetics
   1869 Mendeleev's periodic system of the elements
   1869 Hittorf observes cathode rays in gas discharge
   1873 Maxwell's theory of electromagnetic radiation
   1888 Hertz generates and detects electromagnetic waves
   1895 Lorentz theory of the electron
   1895 Roentgen discovers X-rays
   1896 Thomson and Rutherford discover ionization by X-rays
   1896 Becquerel discovers radioactivity
   1897 Thomson measures charge-to-mass ratio of electron
   1898 Curies discover polonium and radium
c. 1900 Thomson's "plum pudding" model of the atom
   1900 Villard discovers gamma rays
   1900 Planck's constant h
   1901 Roentgen receives first Nobel prize in physics
   1902 Rutherford shows existence of three types of radiation: α, β, γ
   1905 Einstein's special theory of relativity: $E = mc^2$
   1905 Einstein's photoelectric theory: photon energy $E = h\nu$

Historical Outline

- 1907 Thomson proves existence of isotopes experimentally
- 1909 Millikan determines electronic charge accurately
- 1911 Rutherford planetary atomic model
- 1912 von Laue demonstrates interference (wave nature) of X-rays
- 1912 Hess discovers cosmic radiation
- 1913 Soddy proposes name "isotope"
- 1913 Bohr's theory of hydrogen and single-electron ions
- 1913 Coolidge X-ray tube, evacuated tube, and heated cathode
- 1913 Chemical changes occurring as a result of radioactive decay discovered by Fajans and by Soddy
- 1917 Rutherford produces first artificial nuclear reaction
- 1922 Compton effect
- 1924 de Broglie particle wavelength, $\lambda = h/\text{momentum}$
- 1925 International Commission on Radiological Units founded
- 1925 Heisenberg's first paper on quantum mechanics
- 1926 Schroedinger's wave theory of quantum mechanics
- 1927 Heisenberg uncertainty principle
- 1927 Muller discovers that ionizing radiation causes mutations
- 1928 International Commission on Radiological Protection publishes first set of recommendations
- 1928 Adoption of the roentgen as first quantitative unit of exposure
- 1929 Advisory Committee on X-Ray and Radium Protection (now National Council on Radiation Protection and Measurements) organized in the United States
- 1930 Bethe's quantum mechanical theory of stopping power
- 1932 Chadwick discovers neutron
- 1932 Urey discovers deuterium (^2_1H)
- 1932 Anderson discovers positron
- 1934 Curie and Joliot and Fermi produce artificial radioisotopes
- 1936 Gray's formalization of Bragg-Gray principle
- 1938 Hahn observes nuclear fission
- 1942 First nuclear chain reaction
- 1945 First atomic bomb
- 1952 Explosion of first fusion device (hydrogen bomb)
- 1958 Discovery of Van Allen radiation belts
- 1959 Federal Radiation Council created in the United States
- 1960 U.S. Congressional hearings on radiation protection standards
- 1961 First manned orbital space flight
- 1969 Man walks on moon

APPENDIX B

Symbols and Abbreviations

$\propto$	proportional to
$\sim$	approximately, of the order of
°C	degrees centrigrade
μ	micron = 10^{-6} m. Also denotes micro-
μ amp	microampere = 10^{-6} amp
μc	microcurie = 10^{-6} curie
amp	ampere
AMU	atomic mass unit
Å	angstrom unit = 10^{-8} cm
GeV	billion electron volt = 10^9 eV
cal	calorie
cc	cubic centimeter = cm^3
cm	centimeter = 10^{-2} m
coul	coulomb
esu	electrostatic unit (statcoulomb)
eV	electron volt
ft	foot
g	gram
hr	hour
ip	ion pair

Symbols and Abbreviations

j	joule
keV	kiloelectron volt = 10^3 eV
kg	kilogram = 10^3 gm
kVp	kilovolt peak voltage
ln	natural logarithm (base e = 2.713 . . .)
m	meter
ma	milliamp = 10^{-3} amp
mc	millicurie = 10^{-3} curie
MeV	million electron volt = 10^6 eV
min	minute
mm	millimeter = 10^{-3} m
mR	milliroentgen = 10^{-3} R
new	newton
R	roentgen
sec	second
wk	week

APPENDIX C

Physical Constants

Acceleration due to gravity, 9.8 m/sec^2
Avogadro's number, 6.023×10^{23}
charge on electron, -1.602×10^{-19} coul $= -4.803 \times 10^{-10}$ esu
Coulomb-law constant, $K = 8.98 \times 10^9$ new-m^2/coul^2
Mass of electron, 0.000549 AMU $= 9.108 \times 10^{-31}$ kg
Mass of proton, 1.00728 AMU $=$ 1835 electron masses
Mass of neutron, 1.00867 AMU
(For isotope masses see Table 2-1)
Planck's constant, $h = 2\pi\hbar = 6.625 \times 10^{-34}$ j-sec
Speed of light, $c = 2.998 \times 10^8$ m/sec $\cong 3 \times 10^8$ m/sec

APPENDIX D

Conversion Factors

$1 \text{ Å} = 10^{-8} \text{ cm}$
$1 \text{ AMU} = 1.660 \times 10^{-27} \text{ kg} = 931 \text{ MeV}$
$1 \text{ atmosphere} = 760 \text{ torr}$
$1 \text{ barn} = 10^{-24} \text{ cm}^2$
$1 \text{ calorie} = 4.186 \times 10^7 \text{ erg}$
$1 \text{ coul} = 2.998 \times 10^9 \text{ esu}$
$1 \text{ cm} = 0.3937 \text{ in} = 0.03281 \text{ ft}$
$1 \text{ erg} = 10^{-7} \text{ j} = 6.242 \times 10^{11} \text{ eV}$
$1 \text{ eV} = 1.602 \times 10^{-12} \text{ erg}$
$1 \text{ in} = 2.540 \text{ cm}$
$1 \text{ j} = 1 \text{ new-m} = 1 \text{ kg-m}^2/\text{sec}^2$
$1 \text{ kg} = 2.205 \text{ pounds}$
$1 \text{ MeV} = 10^6 \text{ eV} = 1.602 \times 10^{-6} \text{ erg}$
$1 \text{ pound} = 0.4536 \text{ kg} = 453.6 \text{ g}$
$1 \text{ torr} = 1 \text{ mm mercury (pressure)}$
$1 \text{ watt} = 1 \text{ j/sec} = 10^{-6} \text{ megawatt}$

APPENDIX E

Values of the Exponential Function

x	e^x	e^{-x}	x	e^x	e^{-x}
0.00	1.0000	1.0000	2.50	12.182	0.0821
0.05	1.0513	0.9512	2.60	13.464	0.0743
0.10	1.1052	0.9048	2.70	14.880	0.0672
0.15	1.1618	0.8607	2.80	16.445	0.0608
0.20	1.2214	0.8187	2.90	18.174	0.0550
0.25	1.2840	0.7788	3.00	20.086	0.0498
0.30	1.3499	0.7408	3.10	22.198	0.0450
0.35	1.4191	0.7047	3.20	24.533	0.0408
0.40	1.4918	0.6703	3.30	27.113	0.0369
0.45	1.5683	0.6376	3.40	29.964	0.0334
0.50	1.6487	0.6065	3.50	33.115	0.0302
0.55	1.7333	0.5769	3.60	36.598	0.0273
0.60	1.8221	0.5488	3.70	40.447	0.0247
0.65	1.9155	0.5220	3.80	44.701	0.0224
0.70	2.0138	0.4966	3.90	49.402	0.0202

Values of the Exponential Fraction

x	e^x	e^{-x}	x	e^x	e^{-x}
0.75	2.1170	0.4724	4.00	54.598	0.0183
0.80	2.2255	0.4493	4.10	60.340	0.0166
0.85	2.3396	0.4274	4.20	66.686	0.0150
0.90	2.4596	0.4066	4.30	73.700	0.0136
0.95	2.5857	0.3867	4.40	81.451	0.0123
1.00	2.7183	0.3679	4.50	90.017	0.0111
1.10	3.0042	0.3329	4.60	99.484	0.0101
1.20	3.3201	0.3012	4.70	109.95	0.0091
1.30	3.6693	0.2725	4.80	121.51	0.0082
1.40	4.0552	0.2466	4.90	134.29	0.0074
1.50	4.4817	0.2231	5.00	148.41	0.0067
1.60	4.9530	0.2019	5.50	244.69	0.0041
1.70	5.4739	0.1827	6.00	403.43	0.0025
1.80	6.0496	0.1653	6.50	665.14	0.0015
1.90	6.6859	0.1496	7.00	1096.6	0.0009
2.00	7.3891	0.1353	7.50	1808.0	0.0006
2.10	8.1662	0.1225	8.00	2981.0	0.0003
2.20	9.0250	0.1108	8.50	4914.8	0.0002
2.30	9.9742	0.1003	9.00	8103.1	0.0001
2.40	11.023	0.0907	10.00	22026	0.00005

Index

Absorbed dose, 82
Absorption, energy, 69–79
  energy versus intensity, 75, 81–82
  "thin" and "thick" absorbers, 53–54
  also refer to type of particle
Absorption coefficients, 54–58
Accelerator, 135–136
Acute radiation syndrome, 104–106
Agriculture, 145–146
Alpha rays, 21, 22
  cosmic ray, 38
  LET, 96–97
  penetration, 38
  range, 45
  RBE, 97
  spectra, 23, 30–33
Amino acid, 141
Angstrom unit, 9, 155, 158
Angular momentum quantization, 6, 11
Antineutrino, 23, 24, 33
Atomic bomb survivors, 103, 104, 107
Atomic mass (weight), 4, 5, 30
Atomic mass unit, 30
Atomic number, 3, 4, 26
Atomic weight (mass), 4, 5, 30
Atoms, excitation and ionization, 38–40
  structure, 2–5
Auger cascade, 40
Avogadro's number, 5

Background radiation, vii, viii
Bacteria, 145
Bacteriophage, 143
Balance, risk-benefit, 13, 114, 115, 128
Balmer series, 9
Barn, 49
Becquerel, 20, 21, 76

Beta rays, 22, 23, 24
  burns, 108, 110–111
  spectra, 23, 24, 32, 33–34, 38, 71
Biological effects of radiation, 95–115
  acute radiation syndrome, 104–106
  atomic bomb survivors, 103, 104, 107
  beta rays, 108, 110–111
  blood changes, 101, 104–105
  bone marrow, 101
  cataract, 107, 108
  cellular effects, 99–103
  chromosome, 100, 101
  dose-rate dependence, 99, 143–145
  enzyme, 98, 100
  epilation, 101, 105–106
  fatality, 105–106
  gamma rays, 100, 102, 104–106, 109
  genetic effects, 99, 106, 108–112, 143–145
  ingested radioisotopes, 22, 133
  LET, 96–98
  leukemia, 103–104, 107
  life shortening, 107
  mamalian cells, 97
  mutations, 99, 100, 108–112, 143–145
  neutrons, 27, 97–98
  pregnancy, 101, 102, 107, 109
  radioactivity burns, 22, 108, 110–111
  radium dial painters, 22, 107
  sensitivity different cells, 101–103, 114
  somatic effects, 106–108
  sterility, 105
  threshold, 103–104
  virus, 88
  X-rays, 13, 100, 108, 112
Biomedical research, 140–143
Blood, 101, 103, 104–105, 136, 140
Bohr, N., 1, 3, 6–9, 11

Index

Bohr radius, 7
Bone, 103, 107, 138
Bone marrow, 101, 137, 138
   transplant, 105
Bragg, W. H., 83
Bragg peak, 42
Bragg-Gray Principle, 83–84, 85, 87, 89, 126
Brain tumor, 135–136
Bremsstrahlung, 69
Broadening, spectral lines, 16

Calorimetry, 75
Cancer, 107, 133, 134, 136
Carbon dating, 33
Cataracts, 107, 108
Cellular effects, 99–103
Chadwick, J., 26, 27, 61
Chemical detectors, 76
Chromosome, 100, 101
Cloud chamber, 27
Cobalt therapy, 136–137
Collisional broadening, 16
Compton effect, 48, 50–51, 54, 55, 62, 93
Coolidge, W. D., 120, 121
Coolidge tube, 120, 121
Cosmic radiation, vii, viii, 2, 38
Coulombic explosion, 40
Critical organ, 114–115
Crooke's tube, 119
Cross sections, 48–49, 61–67
Curie (unit), 70
Curie, M., 20, 21, 70
Curie, P., 70

Dalton, J., 2
de Broglie, L., 10
de Broglie wavelength, 10, 11
Decay constant, 25
Decay law, 24–26
Decay schemes, 60_{Co}, 70
   226_{Ra}, 32
Depth-dose curves, 91–92
Design, X-ray therapy installation, 128–130
Detectors, *see* Dosimeters and detectors
Deuterium, 4, 29, 30
Diagnosis, 133–139
Diffraction, 10, 118
Disintegration rate, 70
DNA, 100, 142, 143
Doppler broadening, 16

Dose, depth-dose curves, 91–92
   measurement, 83–89, 125–126
   quantities and units, 82
Dose equivalent, 113
Dose rate, 82
   biological effects, 99, 143–145
Dosimeters and detectors, basis, 1
   BF_3 counter, 66
   calorimetry, 75
   chemical, 76
   film, 1, 13, 20, 22, 76, 125, 126
   fluorescence, 1, 12, 13, 20, 22
   gas ionization, 1, 13, 22, 75–79, 125–126
   Geiger-Mueller counter, 75, 78–79, 134, 140
   glass, 125
   luminescent screens, 127
   neutron, 65–67
   pencil in chamber, 126
   proportional counters, 75, 78–79, 88–89
   scintillation, 1, 76, 125, 134
   solid state, 75
   threshold, 65–67
   tissue-equivalent chambers, 87–88
   X-ray, 12–13, 125–126
Dosimetry, 81–93, 125–126
Drift current, 77
Drosophila, 99, 112, 145

Ecology, 146–147
Einstein, A., 5
Electromagnetic radiation, 1–17
   Bremsstrahlung, 69
   Maxwell's theory, 5
   neutron capture, 62
   penetration in matter, 47–58
   quantum theory, 5
   spectrum, 16–17
   velocity, 5
   wave-particle nature, 5–6
Electrometer, 77
Electron, capture by nucleus, 24
   clouds, 11–12
   diffraction, 10
   discovery, 2
   LET, 96
   microscope, 10
   penetration, 38
   photoelectron, 49
   quality factor, 113
   range, 43

Index   163

stopping power, 41
valence, 3
volt, 7, 30
wavelength, 10
Electronic equilibrium, 125–126
Electron microscope, 10
Electron volt, 7, 30
Electroscope, 126
Embryo, 102, 107, 109
Energy absorption, 69–79
Energy-level diagrams, 8, 39
Enzyme, 98, 100, 142
Epilation, 101, 105–106
Epithelium, 101
Erythema, 13, 112
Excited states, 8, 16, 38–40
Exponential function, 159–160
Exponential law, 61
  photon absorption, 54
  radioactive decay, 25
  thermal neutron absorption, 66
  X-ray absorption, 124
Exposure, 82

Failla, G., 81, 87
Fallout, vii, viii, 146
Federal Radiation Council, 114–115
Fermi, E., 132
Fetus, 101, 107, 109
Film, 1, 13, 20, 22, 76, 125, 126
Filters, X-ray, 15, 57, 122
Fission, 27, 62–63, 65–67, 151
Fluence, 69, 72–73
Fluoroscopy, 127
Flux, 55, 72–73
Food preservation, 146
Free radical, 100
Fusion, 29

Gamma rays, 9, 17, 22
  beta decay, 34
  biological effects, 100, 102, 104–106, 109
  60_{Co} decay, 70
  dose, dose rate measurement, 84–86
  interaction and penetration, 47–58
  LET, 96–97
  neutron capture, 62, 91, 148
  quality factor, 113
  226_{Ra} decay, 31–32
  RBE, 97

spectroscopy, 9, 32
Gas amplification, 79
Geiger-Mueller counter, 75, 78–79, 134, 140
Genetics, 99, 106, 108–112, 143–145
GeV, 155
Gram atomic weight, 5
Gray, L. H., 83
Ground state, 8, 38

Half-life, 25, 26, 30
  biological, 133, 134
  effective, 133, 134
  radioactive, 25, 26, 30, 133, 134
Half-value layer, 128
Heisenberg, W., 3, 15
  uncertainty principle, 15–16
Historical outline, 153–154
Hydrogen atom, 3–4, 6–9, 11–12
Hyperthyroidism, 134

Impact parameter, 40, 41
Industrial uses of radiation, 147–150
Intensity, 54, 73–75
Internal emitters, 22, 133
International Commission on Radiological Protection, 112–115, 134
International Commission on Radiological Units, 112–113
Inverse square law, 72
Ion pairs, 74
  recombination, 77
Ionization, 1, 38–40
  continuum, 8, 39
  potential, 8, 39
Ionization chambers, 74, 75, 76–79, 125–126
Isotopes, 3–5, 21, 63, 132–151
  separation, 4

Jackson, H., 120

Kelvin, Lord, 69
K-shell electron capture, 24
kVp voltage, 123

Labeling, 141–143
LD-50, 99, 105
Lea, D. E., 95
LET, LET distributions, biological effects, 96–98
  different radiations, 96–97

measurement, 90
quality factors, 113
RBE, 96–98
relation to stopping power, 82, 96
Leukemia, 103–104, 107, 138, 141
Leukopenia, 105
Life shortening, 107
Lifetimes, excited atomic states, 16
Light quanta, 5
Line breadth, 16
Luminescent screens, 127
Lyman series, 9
Lymphocyte count, 104–105

Mamalian Cells, 97
Mass-energy equivalence, 29, 30
Maximum permissible radiation levels, 113–115
Maxwell, J. C., 5
Metastable states, 16, 24, 134
MeV, 29, 30, 156
Microscopes, electron and optical, 10–11
Mitochondria, 142
Molecules, 2
  dissociation, 39
  rotation-vibration, 9
Monte-Carlo calculations, 90
Muller, H. J., 108
Muscle, 103, 106
Mutations, 99, 100, 108–112
  Dose, dose-rate dependence, 109–110, 143–145
  frequency and type, 110–111
National Committee on Radiation Protection, 112–114
Nerve, 103
Neufeld, J., 90
Neutrino, 23, 24, 34, 63
Neutron, 4, 26–29, 61–67
  activation, 28–29
  atomic weight, 27, 30
  attenuation, 66
  biological effects, 27, 97–98
  capture, 62, 63, 148
  classification, 62
  cross sections, 62–67
  detection, 65–67
  diffraction, 10
  dose, dose rate measurement, 86–89
  fast, 62, 64–65
  fission, 27, 62–63, 65–67, 151
  interaction with matter, 61–67, 73
  nuclear reactions, 61–67
  number in nucleus, 26
  production, 27, 28
  radioisotope production, 63
  RBE, 97–98
  scattering, 64–65
  shielding, 65
  thermal, 62–63, 90
  $1/v$ law, 63
  wave nature, 10
Nuclear power, 150–151
Nuclear reactions, 29–34
  alpha decay, 30–33
  beta decay, 33–34
  deuterium-tritium fusion, 29
  fission, 27, 62–63, 65–67, 151
  neutrons, 61–67
  positron decay, 34
Nucleic acid, 143
Nucleus, atomic, 2–3
Nuisance radiation, 124

Oil exploration, 148
"Over-kill," 97

Pair production, 48, 51–52
Pence, M. L., 119–122
Penetration, charged particle versus photon, 53–54
  charged particle versus photon versus fast neutron, 73
  see Absorption; and refer to type of particle
Periodic system of elements, 3–5
Photoelectric effect, 48, 49–50
Photon, 5
  energy, 5
  interaction and penetration, 47–58
  momentum, 10
Planck's constant, 6, 157
Platelet, 104–105
Population, exposure to radiation, vii, viii, 114–115, 128
Positron, 22, 23, 24, 113
  pair production, 51–52
  spectra, 34
Pressure broadening, 16
Proportional counter, 75, 78–79, 88–89
Proton, 4
  atomic weight, 27, 30
  cosmic ray, 38

range, 44
stopping power, 42

Q-value, 29, 30–34, 63
Quality factors, 113
Quanta, 5
Quantum mechanics, 3, 5–6, 9

rad, 82
Radiation, concentration guides, 115
  cosmic, vii, viii, 2, 38
  detectors, *see* Dosimeters and detectors
  electromagnetic, 1–17
  fallout, vii, viii, 146
  population exposure, vii, viii, 114–115, 128
  protection, 112–115
  protection guides, 114
  sources, vii, viii
  refer also to type of particle
Radioactivity, 20–26
  alpha, 21–23
  beta, 22–23
  decay constant, 25–26
  decay law, 24
  gamma, 22
  half-life, 25–26
  neutron induction, 29–30
  nuclear transformations, 23–24
  positron, 22–24
Radiocarbon dating, 33
Radiocobalt therapy, 136–137
Radiography, 127, 147
Radioisotopes, applications, 132–151
  natural and artificial, 21
  production, 63
  separation, 4
Radium, decay scheme, 31–32
  dial painters, 22, 107
  discovery, 21
  effects in bone, 22
Ranges, 42–45
RBE, 95–99
Reactors, 151
Recombination, 77
Rem, 113
Repair, biological, 104
Reproductive organs, 101
Roentgen (unit), 82
Roentgen, W. C., 1, 12, 13, 20, 47, 118–122

Roentgen-equivalent-man (rem), 113
Roentgen-equivalent-physical (rep), 82
Rossi, H. H., 87
Rutherford, E., 3, 22, 37
Rutherford-Bohr atomic model, 3

Scanning, 137–139
Scattering, absorption coefficient, 55
  Compton, 48, 50–51, 54, 55, 62, 93
  "good" and "poor" geometry, 55–56
  isotopic, 64
  neutron, 64–65
Schroedinger, E., 3
Scintillation, 1, 76, 134
Shielding, medical X-ray installation, 124–125, 128–130
  neutron, 65
  X-ray, 123–125
Snyder, W. S., 91
Solid-state detector, 75
Somatic effects, 106–108
Source spectra, 71
Source strength, 70
Spectroscope, 16
Spectrum, 14–15, 16, 71–72; *refer also* to type of particle
Sterility, 105
Stopping power, 40–42, 69, 82, 96, 135–136

Therapy, 133–139
Thermoelectric generator, 151
Thickness gauge, 148
Thomson, G. P., 118
Thomson, J. J., "Plum pudding" atomic model, 2
Threshold, biological, 103–104
Threshold detectors, 65–67
Tissue rad, 87
Tissue-equivalent chambers, 87–88
Tracers, 133, 141–143, 145–146
Tritium, 4, 29, 30, 141, 143, 144, 146

Uncertainty principle, 15–16

Virus, 98

W-value, 74–76
Waves, 5–6, 11
Work factor, 128, 130

X-rays, 12–15, 118–129

absorption, 13
biological effects, 13, 100, 108, 112
characteristic, 14, 24
continuous, 15
control of exposure, 128–130
detection, 12–13
diagnosis, 13
diffraction, 118
discovery, 1, 12–13
dose, dose rate measurement, 84–86, 125–126
effective energy, 124
exposure rates, 122–123
filter, 15, 57, 122
hardness, 14–15, 121, 124
history, 118–122
interaction and penetration, 47–58
LET, 96–97
maximum photon energy, 15
medical exposures, vii, viii, 13, 115, 128
picture definition, 15
quality factor, 113
radiography, 127, 147
RBE, 97
shielding, 123–125, 128–130
spectra, 14–15, 122, 124
technology, 118–129
tubes, 14, 119–122, 127
useful and nuisance radiation, 124

LIBRARY OF DAVIDSON COLLEGE

Books on regular loan may be checked out for **two weeks**. Books must be presented at the Circulation Desk in order to be renewed.

A fine of **five cents** a day is charged after date due.

Special books are subject to special regulations at the discretion of library staff.

MAY 6 1972						
DEC 9 1972						
JAN 16 1973						
NOV 26 1972						
FEB 20 1981						
OCT 28 '81						